Power Systems

Electrical power has been the technological foundation of industrial societies for many years. Although the systems designed to provide and apply electrical energy have reached a high degree of maturity, unforeseen problems are constantly encountered, necessitating the design of more efficient and reliable systems based on novel technologies. The book series Power Systems is aimed at providing detailed, accurate and sound technical information about these new developments in electrical power engineering. It includes topics on power generation, storage and transmission as well as electrical machines. The monographs and advanced textbooks in this series address researchers, lecturers, industrial engineers and senior students in electrical engineering.

More information about this series at http://www.springer.com/series/4622

G. F. Kovalev · L. M. Lebedeva

Reliability of Power Systems

Springer

G. F. Kovalev
Energy Security Department
Melentiev Energy Systems Institute
of Siberian Branch of the Russian
Academy of Sciences
Irkutsk, Russia

L. M. Lebedeva
Energy Security Department
Melentiev Energy Systems Institute
of Siberian Branch of the Russian
Academy of Sciences
Irkutsk, Russia

ISSN 1612-1287 ISSN 1860-4676 (electronic)
Power Systems
ISBN 978-3-030-18735-4 ISBN 978-3-030-18736-1 (eBook)
https://doi.org/10.1007/978-3-030-18736-1

This Springer imprint is published by the registered company Springer Nature Switzerland AG.
The registered company address is: Gewerbestrasse 11, 6330 Cham, Switzerland

Preface

The monograph includes a description of the meaningful formulation, formalization, and algorithmization of the problem of assessing the reliability of modern, large, and complex electric power systems (EPS) in substantiating their development. It also provides a description of the main factors and operating conditions that affect the reliability of EPS, the assumptions, and simplifications adopted in the respective models. The description and parameters of the corresponding computational tools that solve this problem are also given. Their advantages and disadvantages are discussed.

Being, in the author's opinion, useful for the technical and managerial staff of EPS (energy companies), the published material can be especially useful for young professionals working in the field of ensuring the reliability of EPS and their facilities, as well as for senior students of energy departments prone to scientific research.

Irkutsk, Russia

G. F. Kovalev
L. M. Lebedeva

Introduction

The development of EPS[1] in Russia at the present stage and in the future is characterized by an increase in such quantitative indicators as the installed capacity of generating equipment and the length of transmission lines and, most importantly, by significant qualitative changes. The structure of generating capacities is changing, and network schemes become more complex. There is a unified power system (UPS)—the highest form of the country's energy economy. At the same time, within the UPS, larger and larger territories are covered by a centralized electric system. The UPS of Russia is linked by joint work with the EPS of neighboring countries.

In parallel with this, another process is developing: the process of decentralization of electricity production in the form of distributed generation as sources of electricity of relatively small capacity in the immediate vicinity of consumption. Special attention is paid to sources of local traditional and non-traditional renewable primary energy resources.

Under these conditions, the evaluation of the reliability of large complex EPS is of primary importance.

The dependence of society's life on the electric power industry is such that a reduction in the level of the reliability of EPS leads to significant disruptions in the life of the population, including no less significant industrial damage to the country's public economy. On the other hand, ensuring a high level of the reliability is associated with significant costs.

At the stage of planning and design of EPSs, a high degree of the reliability is put in place. And at the stage of development control, the most significant factors of the reliability of EPS operation concretize:

- provision of energy resources;
- amount, structure, and location of generating capacities, including reserve capacity;
- transfer capability of ties all levels and reserves for them, etc.

[1] Accepted abbreviations and designations are explained in Appendix E.

The high efficiency of the functioning of modern EPS is the result of the fact that the issues of ensuring their reliable work have always been given attention. In the practice of design and operation, rules and computational models for the operation of EPS were developed, taking into account the reliability factor of power supply to consumers. Specialists have accumulated experience in planning the development of EPS, ensuring their necessary reliability.

However, complication of EPSs, the changing conditions of their functioning and development in the market institutional environment requires constant improvement of the existing tools for calculating reliability, upgrading the computational models to determine the quantitative values of the reliability indicators.

The central link of EPS—the basic structure, which includes generating capacities and backbone ties—largely determines the resulting reliability of power supply to consumers. Therefore, ensuring the reliable functioning of the basic structure (system reliability), along with ensuring the necessary reliability of other technological links of EPS (distribution networks, power schemes of specific consumers, etc.), is of great importance.

This operation reflects the results of methodical developments and investigation of the reliability of the basic structure of EPS at the unified (UPS) and integrated electric power system (IPS) levels at the planning and design stage of their development and includes:

1. analysis of practices and solving actual problems of accounting for reliability in controling the development of EPS;
2. the development of a methodic for investigating the reliability of the basic structure in the planning of the development of EPS, taking into account modern requirements of system design practice;
3. mathematical formalization of a meaningful formulation of the problem, the development of algorithms, and their implementation in software complexes for the quantitative evaluation of the reliability indicators of complex systems as the basis for the methodic for investigating the reliability of the basic structure of EPS;
4. conducting experimental investigation on the operability and practical feasibility of developed computer program and approaches for investigating reliability, including ways to improve accuracy and reduce the time required to perform calculations of the reliability indicators;
5. using in the design practice of methodic provisions for accounting for reliability in controling the development of EPS.

When developing the approach of the reliability research, methods of comparative analysis, mathematical modeling, probability theory and combinatorics, mathematical programming, including optimization methods, new computer technologies were used.

Thus, the following is proposed and justified:

- reliability research approaches based on a separate or joint solution using models of two main tasks of the reliability of the basic structure of EPS (selection of the amount and placement of reserves of generating capacity and determination of additional transfer capability of ties), followed by a comprehensive evaluation of the reliability of the structure, as well as adjustment and comparative analysis options for the development of EPS based on such an integrated assessment;
- probabilistic mathematical models of complex EPS, reflecting the main factors and operating conditions of the system, which determine its reliability, such as the failure rate of generating equipment and transmission lines, the configuration of the power consumption schedule, random changes of load, scheduled repairs of equipment, and forms of existence of capacity deficits in the system;
- algorithms for calculating reliability indicators of individual EPS and the system as a whole, reflecting the sufficient accuracy of real processes in them, taking into account the development and possibilities of using modern computer technology and information technologies. When developing the program, attention was paid to ensuring the convenience of its use and a certain universality in the sense of giving the initial data in various forms and varying degrees of accuracy. The program has a modular construction that allows for its further improvement;
- the developed programs used proven effective and accurate algorithms, as well as fundamentally new methods for increasing the accuracy of calculations and reducing computer calculation time. Among them: exclusion from consideration of deficit-free states, minimizing the capacity deficit with a given "policy" of consumers' restrictions, calculating the energy reliability characteristics (ERChs) of ties, and others;
- obtaining dependencies of the sought reliability indicators on the accuracy and form of presentation of the initial information and some calculated characteristics (capacity step, the degree of restriction of the distribution functions of equipment states, etc.) and also analysis of the relationships between different reliability indicators.

In practical terms, the developed method (approach), model, and program allow estimating and correcting the reliability of the main structure of EPS during the design at their upper levels of the hierarchy (EPS, IPS, and regional EPS) and also carry out a technical and economic comparison of the development options of systems, taking into account the reliability criterion. In addition, the influence of a number of factors on the calculated indicators of the reliability of electricity supply was investigated and recommendations were given on the appropriateness and methods of their consideration in practical calculations of the reliability of EPS. So it is shown that the calculated value of reliability indicators is influenced by:

- size and nature of power consumption;
- levels of accident rate of equipment and the degree of its running-in or deterioration;
- size and distribution of generating capacity reserves in the system, accidents and transmission capacity of power lines.

The application of the proposed approach and program can contribute to improving the quality of design, the authenticity of the results obtained, and the degree of validity of the decisions made.

Contents

Chapter 1
The Status of Solving the Issue of Accounting Reliability in Controlling the Development of EPS

1.1 Brief Description of the History of Solving the Problem of Reliability of EPS

It is known that production plants must meet the following criteria:

- of providing the required quantity of products or performance objectives at a high (given) quality;
- of the permissible values of all technical and regime parameters for the conditions of storability and security of equipment;
- of economy and reliability.

Of great importance, among others, is the reliability criterion, since only a sufficient level of reliability for the power plant facilitates the performance of specified functions with the required quality in the field of permissible parameters for a required or long time.

Electric power systems, their separate links and elements, as complex production plants, must also meet the listed criteria, including the reliability criterion.

Attention to the reliability of elements, objects, links, and EPS in general arose naturally simultaneously with the formation and complication of the corresponding levels of the electric power industry. Initially, the problem of reliability, as in any relatively simple technical system, was considered primarily as a problem of failure-free operation, durability, and maintainability of power equipment and was solving, as a rule, on the basis of general engineering considerations, the search for more durable construction materials, improvement of the structures themselves [1]. More important during this period was the provision of the necessary technical parameters and high quality of equipment operation.

The quantitative growth and the possibility of obtaining an economic effect objectively led to the integration of individual energy objects into the system. The formation of EPS is a qualitative leap in the development of the electric power industry, which has set a number of new technical and economic tasks for the specialists.

G. F. Kovalev and L. M. Lebedeva, *Reliability of Power Systems*,
Power Systems, https://doi.org/10.1007/978-3-030-18736-1_1

Among them, the task of ensuring high reliability of the functioning of the systems is being created.

Formation of the first fairly large district power plants both in Russia (the USSR) and abroad took place in the late 1920s—early 30s of the last century. And it is during this period that the works appear on setting the task of ensuring reliable power supply to consumers and the first attempts to solve it [2–9]. In the USSR, the authors of these works were R. A. Ferman, S. A. Kukel'-Krayevsky, R. Yu. Malaya, Ya. M. Tourover, N. S. Afonin, B. M. Yakub, and abroad—G. Klingenberg (Germany), S. A. Smith (USA), and others.

Further development of EPS led to the formation of the IPS, and then, in the 1960s, to the integration of IPSs. As a result, the Unified Power System (UPS) of the country, connected by joint work with EPS of a number of foreign countries, arose.

The reliability problem became topical due to the complication of EPS schemes, large quantitative growth of systems, and qualitative changes in them. The previously developed methods for assessing of the failure-free operation power supply and ways to improve reliability have ceased to "work" because of the increase in the dimension and complexity of tasks. There was a need to form new methods for analysis and synthesis of complex systems, taking into account the reliability factor in control the development and operation of EPS. The importance of the problem is characterized by the number of specialists studying various aspects of the reliability of EPS. A significant scientific, theoretical, and engineering-practical contribution to the solution of the problem was made by V. V. Bolotov, G. A. Volkov, Yu. B. Guk, G. N. Lyalik, I. M. Markovich, M. N. Rozanov, Yu. N. Rudenko, M. A. Syromyatnikov, and others. At the present time in this area continues to operate N. I. Voropai, V. G. Kitushin, N. A. Manov, V. A. Nepomnyashchy, V. R. Hams, Yu. Ya. Chukreev, and others. Foreign ones include R. Billinton, C. Concordia, J. C. Dodu, V. I. Nitu, A. D. Patton, A. J. Wood and others.

As is known [10], the problem of reliability sharply arose in the 1970s and 1980s, not only in EPS, but as an inevitable result of the development of technics in general, and in other manufacturing and technical fields. The universality of the problem led to the birth of a new applied science—the general theory of reliability, the development goal of which was and remains the solution of the most common reliability problems that are relevant to most industry systems.

In this sense, the "General Theory of Reliability" influenced since its inception and will influence the development of methods for investigating the reliability of EPS and their elements. But, on the other hand, the specific nature of EPS does not allow to be based solely on its developments. This is especially true for assessing the reliability of EPS at system levels, which largely predetermines the independent nature of the development of methods for investigating the reliability of EPS.

1.2 Basic Concepts and Definitions Related to the Reliability Problem of Complex Production Systems, Including EPS

In the theory of reliability under the production (technical) system is meant a certain set of elements created by man, interacting in the process of fulfilling the considered range of tasks and interconnected functionally. The presence of significant functional connections between elements distinguishes the system from a simple conglomerate of parts [11]. An element is a system object that performs certain functions in it, for the practical use of which it is necessary to connect the object with other elements into the system.

The concepts expressed by the terms "system" and "element" are methodologically conditional. The same object, depending on the research being conducted, can be considered either as an element or as a system. The object emerges as an element when certain parts of it do not represent a significant interest within the framework of the analysis.

Often, a complex system is advisable to consider as a set of components of its technological links. A technological link is a part of the system consisting of elements combined by performing specified functions that determine the main purpose of this part of the system.

In the research of reliability, in addition to technological (functional) links, systems are often represented as subsystems (regions, nodes). The difference in technological links and subsystems is especially important for EPS, where the technological links are understood as: the generating part of the system, the transmitting part, etc. And the subsystem is the aggregate of all or part of the functional links (generation, conversion, transmission, distribution, and consumption of electricity) united territorially. For example, the UPS, as a system, is the aggregate of its subsystems—IPS, etc.[1]

Methods for investigating the reliability of systems and subsystems, their links and elements depend on such characteristics as complexity, recovery and redundancy.

Complex systems, in contrast to simple systems, in the theory of reliability, it is customary to refer to technical systems that have the ability to rebuild their structure to remain operational in the event of the failure of its elements and parts. The number of possible states of such systems is greater than two. A simple system can be in general only in two states: working and completely non-working.

A complex production system (large system) is usually called a system characterized by the following features: the presence of allocated parts (managed links), participation in the system of people, machines, and the environment, the availability of material, energy and information ties between parts of the system, as well as the ties between the considered and other systems [13].

Large systems, as a rule, are multi-purpose and multi-functional.

[1] The concept of "technological link" is not in [12], but it seems necessary for the authors.

Large systems have a hierarchical structure, assuming the existence of subsystems of different levels of subordination (interconnection).

Multi-purpose system—a system designed to perform two or more tasks. In this respect, EPS has a specific nature, which makes it possible to talk about it as a multi-purpose system, whose goals are to cover the load in each of the energy nodes (districts). For details on this, see Sect. 1.3.

A multi-functional system is a system that has the ability to perform the same task in various ways. This circumstance, as a rule, makes it necessary to conduct appropriate optimization calculations.

Recoverable elements and systems [14] are those that, in the event of a failure, can be repaired during operation, and this recovery is economically justified. The restoration of the object, therefore, is closely related to the maintainability property (see [12]).

A reserve [12] is an increase in the reliability of an object by introducing redundancy. Redundancy—additional elements of the system and the possibilities beyond the minimum required to perform for by the object the specified functions.

The reliability of an object, as a complex property, is largely determined by its unit property, such as failure-free operation [14].

Failure-free operation—the property of an object (system or element) to continuously maintain its operability for a given time or a specified operating time.

Operability—the state of the object, in which it is able to perform the specified functions, keeping the values of the operating parameters within the permissible limits. The fundamental concept of reliability theory—the concept of "failure" is associated with failure-free operation.

According to the State standards [14], it follows that the failure is an event consisting in the violation of operability. Failures have many features, according to which they are subdivided. For the purposes of this paper, the following types of failures are relevant:

- full failure—failure, when the object becomes completely inoperable;
- partial failure—failure, after the occurrence of which the object passes into a state of incomplete operable;
- sudden failure—passage the object to a post-fault state;
- predicted failure—foreseeable malfunction, which makes it possible to prevent it very often by putting into operation other elements or changing the operating mode of the object;
- independent refusal is a refusal, the occurrence of which, unlike the dependent one, is not connected with the appearance of other failures.

The above terms and the most general information do not exhaust the whole set of concepts of the theory of reliability. Here we consider only those that are necessary for the subsequent set out of the issues of reliability of EPS.

1.3 General Issues of Reliability of EPS

EPS as a complex production and technical system is an indissoluble set of functionally connected sources of electricity and transforming (conversion), transmitting, distributing, and regulating devices designed to provide a multitude of consumers with electricity of the required quality in the required quantity.

Realizing that a separate goal is to provide electricity to each particular consumer (in the design EPS scheme this covering of the load of each power node), it makes sense to consider EPS as a multi-purpose system. On the other hand, EPS is a multifunctional system in the sense that the achievement of goals can be realized by implementing various technically permissible modes of loading of generating and network equipment. And at the same time, EPS is a redundant, recoverable, and hierarchical system.

The functioning of EPS is a process of changing the state of the system over time under the influence of external conditions, operating conditions, and internal physical and technical processes, which are mostly stochastic in nature.

The process of functioning of EPS can be represented by a random sequence of intervals of complete operability T_j, alternating with intervals after post-fault states τ_{j+1} ($j = 1, 3, 5, \ldots$), the duration of which is determined by the property of the restorability of EPS and its elements.

The states of completely operable state should be characterized doing of all the specified functions in full volume, but post-failure states (periods of recovery of operability)—both complete and incomplete (partial) doing by the system by the set targets. In this case, the intervals of partially operable and inoperative states τ_j should be supplemented by a characteristic of the depth of failure ΔP_j. The depth of failure is the level of failure of the EPS of its functions. The specific nature of the EPS also requires fixing the place where the system does not perform the specified functions.

The level of development of modern EPS is such that their complete failures are extremely rare events. The overwhelming number of failures is partial. Yes, and in the methodological plan, we can consider that a complete refusal is a special case of partial refusal.

The EPS as a whole can be represented by the following functional links:

- basic structure (generating elements and backbone network);
- distribution networks of the respective energy nodes[2];
- schemes for connecting specific consumers to the energy node.

Figure 1.1 schematically shows the main aspects of the problem of reliability of EPS in general (the upper level) and its links (second level).

Reliability of operation of all links ensures the resultant reliability of EPS in general.

---

[2]Distribution networks of power districts are usually broken down into a number of levels: networks of 220–110 kV; 110–35 kV; 35–6 kV; 0.4 kV.

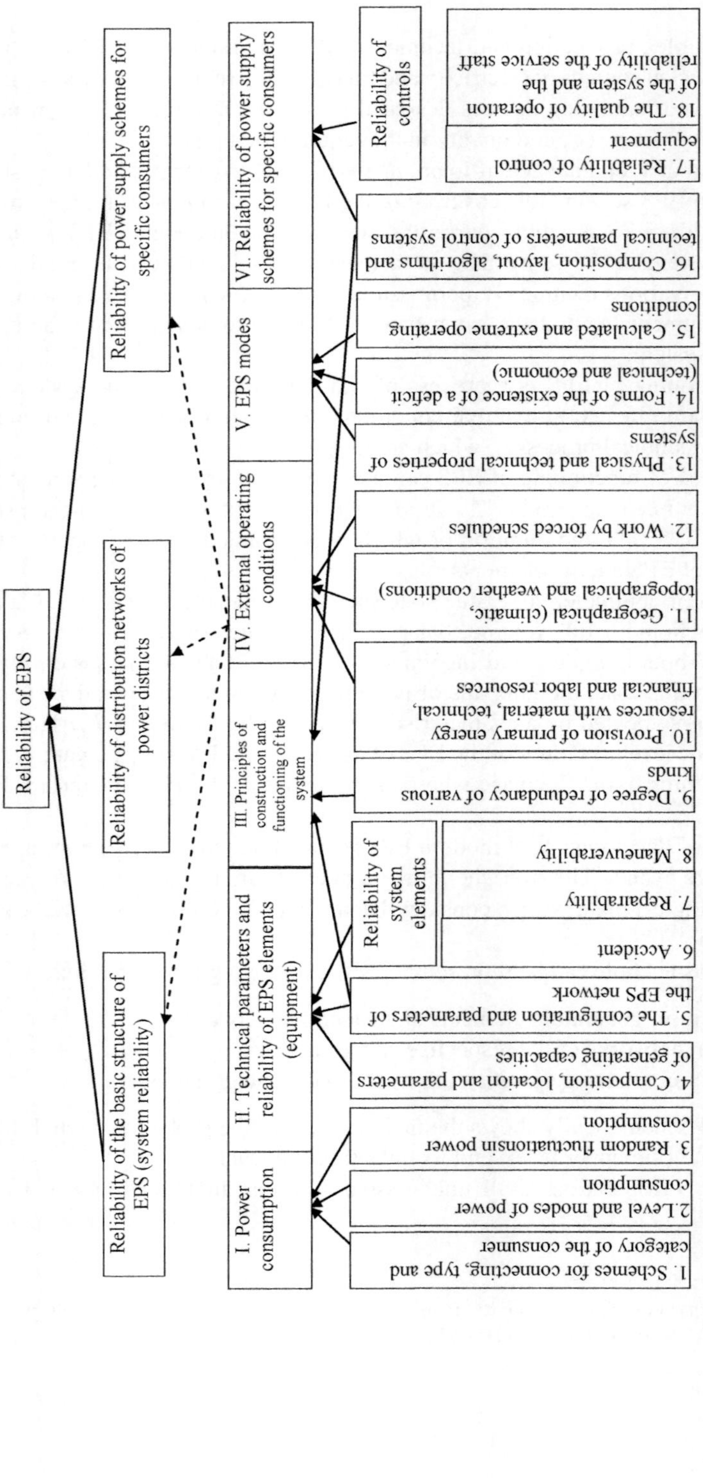

Fig. 1.1 Generalized classification of factors associated with the reliability of EPS

At the third level are the so-called complex factors (groups of factors) that determine the level of reliability of the EPS links. These factors are formed from the sets of individual factors of the operation of EPS.

EPS in general and its links, as well as separate factors, are subjects of research, analysis, and optimization in terms of reliability.

One of the most important links of EPS (see Fig. 1.1) is the basic structure considered in this paper, which provides generation, transformation, and transmission of electrical energy to the regions of power consumption.

Investigation of the reliability of the basic structure (system reliability [12]) involves the consideration of systems at the highest levels: the UPS and the IPS. At the same time, the UPS or IPS is represented by a set of interconnected subsystems (IPS and smaller regions within the EPS, RPS, and RPS regions within the IPS).

Depending on the purpose and the research model used, the elements in the EPS may be either the system's power enterprise (power plants, substations, transmission lines) or equipment (generators, transformers, networks, etc.).

At present, the approach to the problem of reliability has significantly expanded. Previously, reliability was largely understood as the failure-free operation, maintainability, and longevity of the object. Now it is considered that stability, survivability, controllability, etc., are equally important properties of reliability [12].

Hence, it follows that a complete analysis of reliability is a complex analysis of all the above-mentioned unit properties of the reliability of an object. In principle, all these issues should be resolved jointly and mutually connected.

However, so far not only the corresponding models, but even the basic approaches to solving this complex problem are missing, taking into account all the single properties of reliability. Apparently, for a long time, various aspects of reliability will be investigated separately, with the subsequent engineering linkage between them. Accordingly, only problems of reliability analysis are discussed in this work, which are related to the consideration of the properties of failure-free operation and the recoverability of EPS with its reliability indicators.

Regarding EPS in this case, reliability means—the property of the system to fulfill failure-free its main function operation—to supply consumers with electricity of the required quality and in the required quantity.

For EPS, it is generally advisable to consider a complete or partial failure of electricity supply as a failure. In other words, failure-free operation is considered as "deficit-free."

The lack of a deficit in electricity supply, therefore, is the ability of the system to supply by electricity of the required quality at any time to all consumers, regardless of any random changes in the system. Consequently, deficits (limitations) in the electricity supply of consumers, determined by the excess of the required load over the work capacity of the systems, should be considered as failures. This means, any restrictions, explicit or implicit, complete or partial, general or local, caused by any reasons (both internal and external, connected or not related to equipment damage).

The presence of implicit power deficits is determined by the peculiarity of electricity as an EPS product, which consists in the fact that the qualitative characteristics of energy (voltage and frequency) are related to the quantitative characteristic (capacity) of its consumption. In this regard, the capacities deficiency in the system can manifest itself not as disconnection (a cut-off) of some consumers, but as a decrease in the frequency and voltage in the system.

The local nature of the separate capacities' deficiency, even in the case when the working generation capacity exceeds the load of the system as a whole, is also a manifestation of the peculiarity of a complex EPS with constrained transmission capacities on ties.

1.4 The Tasks of Research of Reliability of EPS at Various Stages of Their Management

The complexity of the tasks of controlling the development and operation of EPS is such that it is not possible to make a complex-optimal solution of it on the basis of an all-encompassing formalized ("hard") algorithm. The search for a common solution is known to be carried out in stages, in a consistently and parallel solution of individual problems, linking them to the initial and output information among themselves [15, 16].

In the technical plan (outside the framework of the electricity and capacity market participants), the following territorial-temporal hierarchy of solving the problems of managing the development of EPS has been developed in the design practice [11]:

- territorial levels: UPS, IPS, RES, distribution networks of RES, power supply schemes for individual consumers, power objects of EPS (stations, substations, transmission lines);
- temporary stages: the forecast for development for the future of 15–20 years, the design of development for the five-year period, taking into account the further outlook for 5–10 years, clarification of development plans for the next 1–2 years.

Table 1.1 provides a list of works performed for the relevant territorial levels and time steps [15] in relation to Russian market conditions. One way or another, mutual agreement of these works is carried out. At the same time, one of the stages is supposed to assess the reliable functioning of EPS and their power facilities.

The composition of these works for the relevant territorial levels and time stages from the point of view of the authors should include "reliability problems," that is, problems in which reliability is the main criterion for the solution.

The tasks recommended and formulated in the field of reliability of EPS are given below which can have both an estimated and an optimization nature.

1. Evaluation of the reliability of ensuring the EPS by energy resources.
2. Technical and economic optimization of reliability indicators of EPS equipment for the future.
3. Optimization of the amount, structure, and allocation of reserves of generating capacity.

Table 1.1 List of works performed in the management of the development of EPS

Time stages		Territorial levels					
		Interstate associations	The federal level (UPS)	Interregional level (IPS)	Regional level (RES)	Level of companies and investors	Objective level (stations, substations, ties)
For the state level	Forecasting for 15–25 years	The energy strategy of Russia (in the electric power sector)					
	Planning for a 15 year period	–	General layout of the location of power facilities with details on IPS				
	Planning for up to 7 years	–	The scheme and the development program of the UPS of Russia, including the scheme and the development program of the UNES				
	Planning for a 5–year period	–	–	–	Schemes and programs of long-term development of electric power industry of subjects (regions) of the Russian Federation		
For the level of companies and investors	Forecasting for 10–15 years	–	–	–	–	Corporate strategic program for the development of the generating company	
	Planning for 5–15 years	–	–	–	–	Corporate strategic program for the development of a network company	
	Planning for 5–10 years	–	–	–	–	The company's investment program	
	Plan for 5–10 years	–	–	–	–	–	Business plan of the electric power facility

4. Optimize the additional transmission capacities of links.
5. Optimization of reliability of power supply schemes for specific consumers.
6. Optimization of the reliability power facilities of EPS (stations, substations, transmission lines), taking into account the specific conditions of their operation in the system.

1.5 Indicators, Methods, and Models of Reliability Analysis of EPS

Thus, ensuring the reliability of EPS is a complex and multifaceted problem at different levels of the territorial-temporal and technological hierarchy of management. The main purpose of reliability models is to obtain such reliability indicators that could be used directly or indirectly to make decisions on ensuring the reliability of EPS as a whole and its facilities.

Analyzing reliability, as a complex property of objects (including EPS), it should be considered as a set of unit properties that are relevant to a particular object. This further complicates the problem of analysis and synthesis of reliability of EPS. The creation of a single model, as already noted, is not possible for solving all the tasks of reliability:

- at all time management stages (in forecasting, design, development planning, long-term, short-term and current);
- at all territorial levels (from equipment to unit, facility, and system of varying degrees of unification: regional, unified for the country and above, for the aggregate countries);
- for all technological links—the link of primary energy resources, the generating link, the transport links, the transformation and distribution of electricity (separation of tasks of reliability by technological links is relevant in connection with the replacement of vertically integrated systems in partially horizontally integrated links of the industry as a result of market relations in the electric power industry);
- for all properties unit (security, durability, failure-free operation, maintainability, sustainability, survivability, controllability, resource availability, and storability).

The relation to accuracy and completeness for different levels and stages of solving reliability problems can vary significantly. Moreover, to solve the same problem, it is possible to use models of varying degrees of completeness and accuracy of representation of the initial data and the results of the solution obtained. Thus, the decomposition of a single universal model into a set of models that solve particular problems of reliability is objectively conditioned. But in this case there is a need to harmonize the solutions of particular reliability problems.

Specificity and uniqueness of EPS in most cases do not allow directly using mathematical models and algorithms for calculating reliability indicators, offered in the general theory of reliability. This theory is well developed for a limited class of systems. In these models, the technological features of the electric power systems, their multi-functional and multi-purpose designation, the multiplicity of possible states, and the dominant role of partial failures cannot be fully shown. That is why special mathematical models are being developed to assess the reliability of EPS facilities.

The task of assessing the reliability of EPS is formulated and solved depending on the purpose of the research:

- in advance (for long-term or short-term perspective, with operational control);
- the accepted calculation scheme (depth of equivalenting);
- authenticity and presentation of the source information;
- nomenclature of calculated reliability indicators;
- requirements for the accuracy of the results;
- the used mathematical apparatus.

In domestic and world practice, there are many methodics and approaches for performing calculations of reliability of EPS and its objects [17–21].

Along with a significant number of distinguish in "nuances," they also have certain elements of commonality, the same basic approaches. The variety of such models and the mathematical methods used makes it possible to consider it useful to analyze their features and potentials in order to identify areas of preferential application.

1.5.1 Reliability Indicators

The basis of this or that reliability research approach is usually a quantitative assessment of the level of reliability. Due to the stochastic process of the system functioning, this evaluation represents a probabilistic measure of accomplishment of the tasks facing the EPS, its subsystems, links, or energy associations. The probability measure is represented by different indicators characterizing various aspects of the operation of the object under study.

In the practice of assessing the reliability of EPS, a large number of indicators are used [11, 12, 14, 22–28].

Analysis of the proposed indicators reveals significant differences between them: Some of them characterize the state of the object at a given time, others—at a given time interval; some are relative, others have dimension, etc. But at the same time, most of the indicators have one very significant common feature [1]—they are the same type of operators—the mathematical expectations (m.e.) of random variables. The functions of the distribution of random variables for the evaluation of reliability are rarely used because of the inconvenience of their practical use and poor visibility in comparison with numerical characteristics.

In addition, many of the indicators used are related to each other by elementary relations of the form

$$A = 1 - B,\ E = C - D,\ F = BC = A/D,$$

where A, B, C, E, and F are the corresponding reliability indicators, and D is the received constant value (calculated period, required power generation, etc.). From this, it follows that to reflect the different sides of the reliability of an object, it is necessary to define a limited number of basic reliability indicators.

The used reliability indicators for EPS can be divided into five classes related to frequency of failures, duration of failures, depth of failures, frequency and duration of failures, frequency, duration and depth of failures [11].

For reliability analysis at the system level, the following key indicators are most effective:

(1) technical use ratio (average total relative uptime of failure-free operation) $\mathscr{P}_{\text{tu}}$—mathematical expectation (m.e.) of total duration of T_{Σ} failure-free operation for the period T_{p}, referred to the duration of this period:

$$\mathscr{P}_{\text{tu}} = T_{\Sigma}/T_{\text{p}}$$

Note: Since EPSs as a whole are not deduced for planned repairs, the $\mathscr{P}_{\text{tu}}$ and $\mathscr{P}_{\text{g}}$ indicators (availability ratios[3]) for them are the same $\mathscr{P}_{\text{tu}} = \mathscr{P}_{\text{g}}$.

(2) the average value of the undersupply of electricity W_{und} for a certain period (m.e., of undersupply for a period T_{p});
(3) the relative satisfaction of consumers with electricity π is: in the numerator—mathematical expectation of the amount of electricity actually released, and in the denominator—the required generation of electricity W_{req} for a certain period of T_{p}:

$$\pi = W_{\text{fact}}/W_{\text{req}}.$$

The resulted indicators can be defined both for separate consumers, and for links, subsystems, and EPS in whole.

These indicators in accordance with [12, 14] are complex indicators, because they reflect not one but several properties (failure-free operation, restorability, etc.), reflecting the reliability of the system. These indicators, unlike those of other classes, give the most complete characterization of the reliability of complex objects. However, more accurately characterizing the reliability of the system, they require for their definition much more knowledge about the system under research and a more complete calculating model.

The meaning of determining those or other indicators is to use them to make the necessary decisions when managing EPS at various levels and stages. The numerical values of reliability indicators should fulfill the function of the relevant criteria in

[3] Availability ratios—the probability that the object, being in the waiting mode, will be complete operability at an arbitrary moment of time and, starting from this moment, will work without interruption for a specified time interval.

the process of finding the optimal solution either independently or as part of some more general (universal) criterion.

This criterion is considered by many specialists as a techno-economic criterion of the minimum of the given costs in its most complete form [29–31], which also includes damage from unreliability. The principal form of the criterion in this case is as follows:

$$\min \mathrm{RC} = p\mathrm{CI} + \mathrm{AE} + \mathrm{MD},$$

where RC—the resulted costs; CI—capital investments; p—coefficient of efficiency of capital investments; AE—annual expenses; MD is the mathematical expectation of the accidental value of the damage from the undersupply of electricity during the year due to various types of failures.

The disadvantage of this criterion is the complexity, and in some cases the inability to determine the component MD.

For most consumers, the damage from undersupply is a very complex function of the frequency, duration, depth, suddenness, and time (day and/or of year) the appearance of a capacity deficit. For a number of consumers, as is known, the economic characterization of damage (human sacrifices, etc.) is generally inapplicable.

In these circumstances, the way out of the situation is to standardize the levels of reliability of the functioning of links and EPS as a whole. As noted in [11], "the use of standards in principle does not contradict the above criterion, but, on the basis of it, significantly supplements it and expands, since reliability standards, where necessary, should be economically justified taking into account the accumulated experience of work, analysis of opinions of interested organizations, experimental calculations, etc.".

The availability of standards for reliability indicators greatly facilitates the task of analyzing and synthesizing complex systems, since it becomes possible to formulate variants corresponding to specified standards, with subsequent optimization by the usual criterion of minimum resulted costs (without component damage).

1.5.2 Characteristics of Assessment Models (Synthesis) of Reliability of Electric Power Objects (Systems)

Significant specificity and uniqueness of EPS in most cases, as already noted at the end of Sect. 1.1, do not allow us to directly use mathematical models and algorithms for calculating reliability indicators currently proposed in the general theory of reliability. This theory is well developed for the so-called simple systems and a limited class of complex systems, in whose models the technological features of EPS cannot be sufficiently fully reflected, their multi-functionality and multi-purpose purpose, the multiplicity of probable states, and the dominant role of partial failures. That is why to evaluate the reliability of EPS, it is necessary to develop special mathematical models, which are then implemented in algorithms and calculation programs. Analysis of the features and potentialities of the set of such models and the mathematical methods used will allow us to identify areas of their preferred application and extend the reliability assessment through its different indicators.

The task of assessing the reliability of EPS at different territorial and temporal levels of management according to the approach used is put and solved in different ways depending on the purpose of the study, the limitations on the decision time, the calculation scheme, the authenticity and presentation of the source information, the nomenclature of the calculated reliability indicators, the accuracy requirements for the results obtained, and the used "mathematical apparatus." Therefore, there are many different models for evaluating the reliability of EPSs.

However, all the diversity of these models in the methodological terms can be classified according to the nature of their use (see Table 1.2).

When solving the tasks of the EPS management taking into account the reliability (in terms of the structure of generating capacities, the structure of electrical ties, the choice of parameters and means of controlling the system, etc.), reliability indices are calculated either by a separate (independent) reliability assessment model, or in the process of solving of the main problem (algorithmically together, as in [15, 16, 22, 23, 31–35]).

The presence of independent models of reliability evaluation (synthesis) allows solving the individual problems of managing the development and operation of EPS to be limited to rough preliminary consideration of reliability (up to the application of mediated norms and "reliability rules") with subsequent in-depth analysis from the point of view of the ensured reliability.

Refinement of the values of reliability indicators is performed only for the identified best variants calculated using a special model for reliability assessment, taking into account more factors affecting reliability. If you need to calculate reliability indicators with great accuracy, this path as a whole requires less effort and less time. A significant role is assigned to independent models in the field of research on the reliability of EPS. At the same time, they are used for both a comparative analysis of the options, and also in two aspects, reflected in Table 1.2. Such researches make it possible to formulate more reasonably new models of reliability assessment, including those that would provide practically the required completeness and accuracy of calculations with an increased speed of the algorithm due to possible simplifications.

The application of models for assessing the reliability of EPSs as subblocks in the program for optimizing the development of EPS allows to exclude an additional estimate of the reliability of the resulting solution. In this case, of course, the model must have satisfactory speed and accuracy, be sufficiently complete. The periodicity of the operation of such a model in the framework of a computer complex, at each iteration or at the conclusion of the formation of some variant or step of the calculation, will be determined by specific models, but, apparently, periodic operation of the subblock is expedient for speed reasons.

Table 1.2 Classification of models of reliability of EPS and their objects (at the level of qualitative estimates)

Model usage characteristics	The degree of completeness of the account of influencing factors	Accuracy of presentation of factors considered	The speed of the corresponding programs on the computer	The purpose of using the model
Independent model of reliability	Practically complete accounting	The most accurate	Low or medium	Investigation of the reliability properties of EPS and their objects, as well as the properties of the corresponding models
	Enough full accounting	Quite accurate	Average	Comparative analysis of reliability of control object synthesis options
				Optimization of reliability of EPS and their facilities
The reliability model as part of the optimization model for solving a particular control problem	Enough full accounting	Approximate	High enough	Estimation of reliability in the process of optimization calculation
	Insufficient accounting	Approximate, even rude	High	Approximate reliability accounting in the optimization model

1.5.3 Characteristics of the Main Methods Used in Models for Assessing the Reliability of Facilities and Systems of Electric Power Industry

The methods on which models for evaluating the reliability of EPS and their objects are based are known in the general theory of reliability [13, 28, 36–38], but their application in this case differs in some features (Table 1.3).

For the developed methods and models of reliability research, a large number of different methods are used.

Table 1.3 Classification of methods for determining indicators of reliability of EPS and their objects

Group	Class	Kind			Area of primary use
A. Methods for the experimental evaluation of reliability	I. Tests on the reliability of a real object	1. Long-term testing			Determination and verification of reliability of EPS equipment
		2. Accelerated tests			
B. Methods for the calculation of reliability	II. Methods that do not require element-wise modeling of an object	3. Retrospective methods			Short-term planning of the work of EPS
		4. Methods of extrapolation			Perspective and long-term planning of the development of EPS and their facilities
		5. Expert methods			
	III. Methods based on element-to-element modeling of a research object	6. Deterministic methods	6.1. Methods of physical modeling		At all territorial levels and temporary stages of controlling the development and operation of EPS as a primary or auxiliary means
			6.2. Criterion $n - i$		
			6.3. Worst-Case Method		
			6.4. Standardization		
		7. Probabilistic methods	7.1. Analytical methods based on the representation of real stochastic phenomena	(a) By random events	
				(b) By random processes	
			7.2. Statistical methods based on the representation of real stochastic phenomena	(a) By random events	
				(b) By random processes	

When solving any problem, including the task of assessing the reliability of EPS, the preference for any method should be determined by the content of the problem being solved. In each case, the goal is to obtain a sufficiently fast and convenient computational program that gives satisfactory results.

The variety of methods can be visually classified by some principal features and, first of all, from the point of view of providing information for the process of determining reliability indicators and the used mathematical software.

It is proposed to distinguish between the following two groups of methods: an experimental estimation of reliability (group A), and calculations and prediction of reliability (group B, Table 1.3, first column).

The methods of experimental reliability assessment (group A) are based on research of the results of special tests ("tests on reliability") conducted on the objects themselves or their physical models. Special tests are understood as the process of determining or checking the reliability indicators by empirical testing. The main purpose of such tests is to create an information base (obtaining more complete and authenticity data about the object as a physical reality, however, the availability of this data is still limited). Experimental definition and verification of reliability indicators are possible at all stages of management: design, manufacture, and operation, but not for all objects. Such a database can be used for calculation methods of analysis and reliability analysis or for industrial control of the level of reliability of industrial products.

A special place in this group is occupied by methods for research in the process of testing the physicochemical and other causes of failures that require the conduct of multi-purpose experiments of the corresponding nature.

Methods for the calculation of reliability (group B) are used to determine the numerical characteristics of the reliability of the research object under the known structure, operating conditions, and reliability indicators of its constituent elements.

Methods, depending on the methodological principles and the mathematical apparatus used, are divided into classes (Table 1.3, second column):

I. Tests for reliability are divided into long-term and accelerated (Table 1.3, the second and third columns). The main principle of long-term tests is the reproduction of the real operating conditions of the facility. Accelerated tests are characterized by forced loading modes of the object in order to obtain the necessary information on reliability in the shortest possible time (in comparison with the operating conditions).

II. Methods that do not require detailed (element-to-element) modeling of the object:

- Retrospective—based on the generalization of past experience;
- Extrapolation—based on analysis and forecasting of current trends;
- Expert—based on the knowledge, experience, and intuition of specialists.

These methods are used in forecasting to estimate the numerical values of reliability indicators of an object under conditions of incomplete certainty as the quantitative characteristics of the reliability of the elements making up the object and the conditions for its functioning.

III. Methods based on element-to-element modeling of the object of research. These methods are subdivided into so-called deterministic and probabilistic ones. Among the probabilistic methods, on the one hand, they distinguish: "analytical," based on the use of: (1) functional relationships in the form of mathematical dependences, (2) analytical expressions of probabilistic processes; (3) full or truncated use of all possible states of the object; and "statistical" using the apparatus of the Monte Carlo method or pseudo-statistical methods such as LP_τ programming [36].

On the other hand, probabilistic methods can be based on the representation of stochastic phenomena by random events or random processes. This class also includes physical modeling methods, which often require an element-to-element representation of a complex object.

Next, we give a brief description of the methods from the point of view of their division into species and the definition of the area of preferential use (Table 1.3, third and fourth columns).

Group A methods are not widely used for comprehensible reasons (uniqueness and other features of EPS). Special tests for reliability require either very long tests of a small number of elements or a very large number of equipment (objects) with short tests. For most power facilities, neither is unacceptable for economic reasons, and because of the uniqueness and for most power plants, neither is unacceptable for economic reasons, and because of the uniqueness and the single copy of certain types of equipment of certain types of equipment. To identify and eliminate structural (constructive) defects in the electric power industry, bench tests of one or two head units have been adopted, but not to assess reliability. This, of course, increases reliability, but does not give estimates of reliability indicators.

However, the method of long-term testing is used in the electric power industry in its extreme form—in the form of analysis of operational experience (statistical data) of real objects (systems) and their equipment. For this reason, information systems and reliability services should be formed in the EPS. Large and labor-intensive work should be carried out to collect, process, store, and use data on the reliability of electric power plants.

The use of accelerated testing is limited in the electric power industry by the level of uncomplicated equipment and individual units and elements of more sophisticated equipment.

Group B methods are preferably used in models for assessing the reliability of EPS and their objects due to their greater generality, rigor, and less dependence on subjective factors. These methods are used at all territorial-temporal levels as a primary or auxiliary means.

Methods that do not require element-to-element modeling of the object of research are subdivided into the following types (Table 1.3, item II (3, 4, 5)):

II.3. Retrospective methods are calculated methods for assessing reliability, based on an analysis of past experience of the operation of the facility and the justified use of this experience under the forecasted conditions for its development. Most often, the method of planning the experiment and regression analysis is used.

The application of these methods is possible under the condition of certain "inertia" in the development of the object, that is, sufficient stability of its properties and structure over time, which in general is characteristic of EPS and their objects. An advantageous field of application of these methods is the analysis of the functioning of EPS or their individual elements in the long-term and perspective planning of their development.

II.4. Extrapolation methods are methods for predicting reliability based on an analysis of trends of changing the reliability of an object from changing its individual parameters or working conditions. These methods, widely used to predict the reliability of EPS equipment, are sometimes also used to assess the reliability of objects, subsystems, and the power system as a whole. These methods assume the use of representative statistics collected for a sufficiently long period of time or for a large number of objects. But since EPSs as objects of research change in time in a complex way, there are problems with the accuracy of the results obtained, especially when assessing the reliability of EPS facilities for a long-term perspective.

II.5. Expert methods are methods of assessing reliability based on the knowledge, experience, and intuition of specialists accumulated in this field. These methods are advisable to use in those cases when the process of determining reliability is not formalized, or data on the development of the object are essentially undefined.

Methods based on element-by-element modeling of the research object are divided into two main types: deterministic and probabilistic (Table 1.3, points III, 6, 7).

III.6. Deterministic are methods and criteria for the analysis and synthesis of the reliability of an object (system) in which the probabilistic characteristics of failures of the elements of the object are not modeled, but only the ability of the object to withstand any perturbations from a priori of a given class is analyzed, that is, the functioning of the object after such a perturbation must satisfy given conditions and parameters. For EPS and their facilities—this is the permissible levels of voltage, frequency, equipment load, deficit-free of capacity and energy. From the deterministic methods, the following are distinguished.

III.6.1. Physical modeling methods are, in principle, applicable to the experimental evaluation of reliability, especially at the system level, but this possibility is still used to research individual states or regimes of the system in deterministic form, for example, in the research of survivability. And it is unlikely to be widely used for probabilistic analysis because of the complexity of conducting such experiments.

III.6.2. Reliability of the object in the event of failure of any i elements (reliability by criterion (rule $(\boldsymbol{n}-\boldsymbol{i})$) is the property of the object to perform basic functions in case of failure $i = 1, 2, 3 \ldots$ elements of n even under unfavorable, but really possible conditions. In this case, the load of all elements must remain within the permissible limits, the regime parameters should not exceed the standard ranges; however, deterioration of the effective operation of the facility is generally allowed (fuel consumption, energy loss, etc.).

II.6.3. The "worst-case" method assumes that the object must perform its functions when the composition and parameters of its elements, taking into account the surrounding conditions, have limiting values.

III.6.4. Standardization involves taking into account the reliability factor by applying the specified rules and reliability indicators for the structure and parameters of the object.

III.7. Probabilistic methods and criteria are used to estimate how often and how much time an object (system) will be in an inoperative state due to failures of elements. For such an assessment, it is mandatory to model probabilistic processes in the system and the probability characteristics of element failures. It is believed that the probabilistic approaches give more versatile, deep, and accurate characteristics of the reliability of the object in comparison with the deterministic, but they are more complex and time-consuming.

III.7.1. In analytical methods of calculating reliability, as a rule, the main provisions of probability theory, combinatorics, algebra of logic, queuing theory, etc., are used.

Analytical methods in the presence of a mathematical description of functional relationships between individual factors fundamentally allow to solve any task of assessing reliability in the electric power industry with the necessary accuracy. In practical terms, the absolute advantage of these methods over others is hampered by: the absence or bulkiness in some cases of describing functional relationships; "curse of dimension," making it impossible in some cases to perform calculations even on modern computers in an acceptable time; difficulties associated with the calculation of certain indicators.

- The method of structural schemes ("block scheme"), using a conditional graphic image of the elements of the system and the connections between them (the structural scheme), while the calculation of reliability indicators is made with the help of graph theory or by means of sequential equivalenting;
- a phase space method that examines the random process of functioning of a complex object on the phase space—a set of states that differ in the composition of the serviceable and faulty elements;
- Markov processes method (state space method) is a special case of the phase space method in which the process of object functioning is described by means of a system of differential equations of state transitions from state to state;
- network method (the method of minimum paths and cross sections);
- topological methods, etc.

In the field of statistical methods, one distinguishes:

- *Functional-statistical method*, in which the process of the functioning of the system is described by a probabilistic model, repeatedly tested by means of computer programs;
- *logical-statistical method*, in which the structure of the system and the features of its functioning are described by means of algebra of logic, and reliability calculation is carried out with the help of statistical modeling, etc.

Methods for calculating reliability are also different depending on whether or not the recoverability of equipment is taken into account, only the moments of the appearance of failures are taken into account, or the states of the object in failures are taken into account, different ways to optimize of states (deficit modes for EPS), etc.

Note that given in Table 1.3 the classification of methods in a certain sense is formally abstract. In practice, in the same model, one can find a combination of different methods, each of which is used to solve private subtasks. As experience shows, it is the combined models that prove to be the most acceptable from the point of view of the required accuracy and speed of calculate.

1.5.4 Characteristic of the Generalized Model of the Evaluation of the Complex Property of Reliability

The development of models and methods for investigating the reliability of EPS comes from assessing reliability in terms of failure-free operation and maintainability (restorability) to the investigate and evaluation of other unit reliability properties.

It should be noted that when creating models, *the unit properties of reliability* that are topical for EPS have a different degree of significance. Therefore, in their study, it is important to follow a certain sequence.

For functioning systems, this is not so important and each property can be evaluated separately from the others. But when designing, the sequence of ensuring unit properties is principal. For example, it is pointless to engage in ensuring high survivability without providing the necessary level of safety, durability, failure-free operation, maintainability, and sustainability.

It is known that:

- unit properties are interrelated and interdependent;
- the relative cost of providing the required level of different unit properties is different;
- in a properly ranked sequence, the ensuring of each subsequent unit property is cheaper if all the previous properties are ensured.

Given these circumstances, a rational sequence for the investigation of unit properties of reliability of EPS should be as follows:

1. Security.
2. Durability.
3. Failure-free operation.
4. Maintainability.
5. Stabilability.
6. Survivability.
7. Controllability.
8. Resource availability.
9. Storability.

The temporal aspect of reliability is represented by such concepts as balance reliability, regime reliability, current (switching) reliability, and so on. For them, the models and methods discussed above are also applicable, but taking into account the specifics of the problem being solved.

Reliability, estimated by this or that model, is a "calculated reliability," which cannot exactly coincide with the actual reliability of the object, but only approximately correspond to it because of the assumptions and simplifications introduced into the model. Therefore, from the set of models, one should be chosen that could be assigned the status of "normative models." The use of such models for appropriate calculations would provide a comparison of the reliability of objects on a regulatory ("legalized" basis).

In world practice, for example, such well-known programs as MEXICO (France), SICRET (Italy), COMREL (Canada), TRELSS (Poland), CONFGT (Brazil), etc.

The problems of modeling the reliability of EPSs in our country were paid attention to Volkov G. A., Gol'denberg F. D., Guk Yu. B., Lyalik G. N., Malkin P. A., Oboskalov V. P., Rozanov M. N., Rudenko Yu. N., Fokin Yu. A., Chukreev Yu. Ya., and others [21, 22, 39–42].

Abroad—Billinton R., Concordia, Dodu I., Allan R., Salvadori L., Endreni J., and others [18–20, 43].

The model for estimating the reliability of an object (system) in the general case consists of three main blocks:

1. **Probabilistic block** for the formation of the calculated states of an object.
2. **Block of optimization** of operating modes of the object in the calculated states, determined at the first stage.
3. **Definition of indicators of reliability** of the object from the data obtained in the first two stages.

All the above methods are applicable for the **first stage** of solving the problem of reliability assessment.

At the **second stage**—in the block of optimization of the calculated states (modes) in the organizational and commercial structure, in addition to the technical requirements, it is necessary to take into account the fulfillment of contractual obligations of various interrelated technological links of the system (production, transmission, distribution). The evaluation of reliability in the system must take into account market relations in the electric power industry and, therefore, it is required to determine the relevant reliability indicators. This block is the central unit of the reliability assessment model. It is subject to increased requirements in terms of completeness of accounting for the main factors, accuracy and speed.

The **third stage**—the processing of the accumulated results of the first two stages is not particularly difficult.

Let us consider in more detail optimization of reliability of complex EPS. By complex, as already noted, here are meant EPS, the design schemes of which consist of a number of energy nodes and connections between them with limited capacity. Energy nodes are represented by concentrated subsystems with generalized load and generating equipment, and the internal network of which does not interfere with the transfer of power from the sources of electric power generation to the load buses in all possible modes (the modern name for concentrated subsystems is the free flow zones). Complex EPSs (especially, e.g., the UPS of Russia or even interstate associations) allow achieving a certain economic effect through mutual assistance (in comparison with the conditions of their isolated work). The realization of this mutual assistance is carried out due to the creation of intersystem ties of sufficient communication capacity; therefore, the optimal value of the generator power in the association and its placement between the EPSs are sought together with the definition of communication capacity.

The calculation model for optimizing the variant of the development of a large complex EPS by the reliability factor is intended for selecting the basic structure of EPS and should include, along with optimization of generating capacities and transmission capacities of ties, also the link of provision EPS by primary energy resources. In general, restrictions on all types of resources (material, labor, financial) should be taken into account.

The model should provide for two criteria for calculating reliability, both in the form of technical and economic minimization of the resulted costs for reserving of capacity, primary energy resources, and networks, taking into account compensation for likely economic damage from unreliable electricity supply to consumers, and in the form of ensuring a standard of reliability.

Technical and economic optimization of reliability should be based on the following provisions.

The optimum level of reliability of multi-node EPS corresponds to the minimum of the total resulted costs:

$$Z_{\Sigma} = Z^{\mathrm{G}} + Z^{\mathrm{TL}} + Z^{\mathrm{E}} + D, \tag{1.1}$$

where

$Z^{\mathrm{G}} = \sum_{m=1}^{M} z_m^{\mathrm{g}} \cdot R_m^{\mathrm{G}}$, $Z^{\mathrm{TL}} = \sum_{n=1}^{N} z_n^{\mathrm{tl}} \cdot R_n^{\mathrm{TL}}$, $Z^{\mathrm{E}} = \sum_{m=1}^{M} z_m^{\mathrm{e}} \cdot R_m^{\mathrm{E}}$—costs, respectively, in the reserve of generating capacity in all M nodes of the system, in increasing the capacity of all N system ties and in reserves primary energy resources; Z^{G}, Z^{TL}, Z^{E}—unit costs in the reserve of generating capacity of the m-th node, in increasing the capacity of n-th tie and in reserve of primary energy resources;

R_m^G, R_n^{TL}, R_m^E—accordingly, the value of the reserve of generating capacity in the m-th node, the additional capacity of n-th tie, the reserves of primary energy resources;
$D = \sum_{m=1}^{M} d_{0m} W_m^{und}\left(R_m^G, R_n^{TL}, R_m^E\right)$, $m = \overline{1, M}$, $n = \overline{1, N}$—total costs for the system as a whole to compensate for economic damage from the average annual undersupply of electricity to consumers in all M nodes of the system;
y_{0m}—the specific value of the averaged damage from interruptions of power supply to consumers of the m-th node;
W_m^{und}—the mathematical expectation (m.e.) of undersupply of electricity to consumers of the m-th node, depending on the levels of reserve R_m^G, R_n^{TL}, R_m^E in the system.

The conditions for the minimum of the functional (1.1) in this case are equality to zero of partial derivatives with respect to the means of ensuring reliability for individual nodes R_m^G, R_m^E and R_n^{TL} ties.

$$\begin{aligned}
\frac{\partial Z_\Sigma}{\partial R_m^G} &= \frac{\partial Z^G}{\partial R_m^G} + \frac{\partial D}{\partial R_m^G} = z_m^g + \frac{\partial D}{\partial R_m^G} = 0, \quad m = \overline{1, M};\\
\frac{\partial Z_\Sigma}{\partial R_m^E} &= \frac{\partial Z^E}{\partial R_m^E} + \frac{\partial D}{\partial R_m^E} = z_m^e + \frac{\partial D}{\partial R_m^E} = 0, \quad m = \overline{1, M};\\
\frac{\partial Z_\Sigma}{\partial R_n^{TL}} &= \frac{\partial Z^{TL}}{\partial R_n^{TL}} + \frac{\partial D}{\partial R_n^{TL}} = z_n^{tl} + \frac{\partial J}{\partial R_n^{TL}} = 0, \quad n = \overline{1, N}.
\end{aligned} \tag{1.2}$$

The purpose of the calculations is to obtain values R_m^G, R_m^E and R_n^T, for all nodes and ties, respectively, satisfying the conditions (1.2).

When optimizing calculations for given (standard) reliability indicators, the optimality criterion looks like this.

The optimum level of reliability meets the minimum cost

$$Z_\Sigma = Z^G + Z^{TL} + Z^E \Rightarrow \min \tag{1.3}$$

when

- $IR_m = IR_{standard\,m}$, $m = \overline{1, M}$ or $IR_m \to \max$ $m = \overline{1, M}$ with limited resources,
- the provision of $IR_m = IR_{standard\,m}$, $m = \overline{1, M}$ is not possible.

That is, when the problem of the optimal distribution of limited resources is solve.

Here, IR is a calculated indicator of reliability, $IR_{standard\,m}$ is the value of standard of this indicator in the m-th node.

As in the first case (assessment of damage from undersupply of electricity), and in the second (direct use of the standard index of reliability), the calculation of reliability indicators as a subblock is included in the optimization model of reliability of EPS.

In addition to specialized optimization programs containing a reliability evaluation unit, independent reliability models can also be used for optimization purposes in an interactive mode. In a number of cases, this gives a positive effect (the number of iterations decreases, the accuracy of determining reliability indicators increases).

The evaluation models used in the framework of optimization programs, at the present stage, must satisfy the requirements set forth below.

The evaluation computational model for calculation of reliability indicators of EPS is recommended to be used in cases when temporary or resource constraints do not fully provide economically reasonable standards of reserve, and the choice of means of ensuring reliability is so narrow that it is possible to form only a few alternative variants with the help of an expert. In such cases, using the evaluation model, a technical analysis of the options is carried out. By the results of the analysis, reliability indicators are calculated and they are compared with the standards. Preference, as a rule, is given to a variant with a higher level of reliability and/or lower costs.

A specialized evaluation model should provide calculation of reliability characteristics of complex EPSs that can be represented by any one, including a multi-ring calculated scheme and with constrained transfer capabilities of ties among the nodes. In this model, current and capital (average) repairs should be calculated. The calculation modes formed in the model itself should be optimized.

The results of calculations for such a model should be:

(a) reliability indicators;
(b) the values of the estimated reserves of various types;
(c) the values of transfer capabilities of ties for ensuring the mutual reservation of EPS and the load distribution function of these links; the values of the required primary energy resources (water at HPP and fuel at TPP);
(d) assessing the deficiency of key resources (generating capacity in nodes and d) assessing the deficiency of key resources (generating capacity in nodes and transfer capabilities of ties) to ensure reliability to ensure reliability.

The information obtained should provide opportunities for a deep engineering analysis of the calculated reliability as the properties of a particular EPS in the specific conditions under consideration and allow to detect the bottlenecks in the EES in terms of reliability.

Since during the work of the evaluation program all possible states (modes) of the system are viewed, the characteristics of the flow over the ties obtained as a result of calculations are sufficiently complete information on their loading and do not require specification by analyzing the calculated of long and maximum states (modes), as is done in simplified approaches.

1.5.5 The Main Stages of Creating a Model for Assessing the Reliability of Electric Power Systems

Investigation of the reliability of EPS is a typical problem considered in the theory investigation of operations [44]; therefore, here (we recall), as in any development of system, the following stages are distinguished:

(1) statement of the research task in its technological form (meaningful formulation of the problem);
(2) construction of a mathematical model of the object under research, an algorithm, and a program for calculating the desired quantitative indicators;
(3) conducting test calculations;
(4) practical evaluation of the solutions obtained, testing the working capacity of the model;
(5) formulating and introducing into the practice management of object the methodic of the developed investigation.

This cycle of research, the characteristic feature of which is the interconnectedness of its constituent stages, can be repeated many times in order to gradually improve the methodic for solving the problem of the transition from analysis to synthesis of the system. The peculiarity of the construction of cycles is the moment of mandatory practical verification of the created methodic. As a result of comparison of the solutions obtained with the experimental data (the process of validation and verification of the model), the possibilities for correcting and improving the model and methodic are revealed.

When setting the task of research of reliability of EPS, it is necessary to pay attention to the following circumstances:

- Since there is a territorially-temporal and functional hierarchy of system control and a corresponding division of the global management task into a number of particular tasks, the setting of each such task should be carried out taking into account the control systems of EPS, and the practical implementation of which is possible. Otherwise, it is possible to develop a methodic that is not suitable for practical purposes;
- The boundaries of the selected task, the admissibility of the physical and technical processes taking place in the system, as well as the requirements for the initial data and the sought quantities, must be clearly defined.

The development of a computational model (sometimes called a "reliable model" [37]) isomorphic to a complex system is an unreal thing, so any model will always be approximate. Despite this, the use of computational models is the only possible way to study the reliability of complex systems, to which EPS belong, and, above all, at the stage of designing their development. In [38], it is emphasized that the more complex the system under investigation, the more effective, despite their approximations, is the use of mathematical calculation methods at all stages of its design and operation.

To form a model, in addition to taking into account the specifics of the problem being solved, a clear definition of the goals of creating a model and the choice of mathematical methods of solution, it is also required to choose the calculated quantitative indicators and the criteria for optimality of the solution of the problem (see Sect. 1.5.1).

The decision-making procedure for the control of EPS with respect to the reliability factor, like any procedure for making a rather complex decision, is most often in considering several possible and feasible alternatives and choosing the best alternative. The ways to optimize the solution may be different and depend on the elaboration of the methods of direct synthesis of a particular link of EPS with respect to reliability. As is known, direct synthesis is a complex and not always achievable task, which is currently the case with EPS.

The creation of direct synthesis methods will represent the final stage of research into and analysis of systems reliability. In [45], it is proposed and accepted, and up to now methods of comparative analysis of systems reliability are used, which make it possible to select the best from the series of variants according to the accepted criteria. Effective use of comparative analysis is possible and feasible even in conditions of a simplified approach for determining indicators of reliability, inaccuracy, and uncertainty of the initial information and the lack of standards. An important feature of this method is the possibility of using not the absolute values of the calculated indicators, but the relationships between them.

Thus, the conditions of comparative analysis allow us to calculate the quantitative values of reliability indicators with certain assumptions and errors that are equally apparent in all the compared options. This is especially important, because at project levels information about the equipment failure rate, load levels, etc., which primarily determines the reliability of power supply, can be considered only known approximately.

Summarizing the content of this section, we can draw the following conclusions, which were the basis for the research presented here.

1. Among the available methods of analyzing the reliability of EPS at the system level, the most expedient are the calculation methods (methods of mathematical modeling).
2. The creation of a sufficiently authenticity specialized "reliable" model of EPS is an indispensable stage in the organization of research on the reliability of EPS.
3. Creation of a reliable model is a multi-stage process, the important moments of which are the practical verification of the results obtained with the help of the model and its improvement as a consequence of these checks and new methodical developments.
4. Different control tasks require different models, including reliability models. This, in particular, explains the objective necessity of their great diversity.
5. Some justified simplifications and assumptions in the reliability models are unavoidable and cannot serve as an obstacle to the comparative evaluation of the investigated variants of the solution of the corresponding control problem from the point of view of reliability.

6. Since the comparison of the efficiency of options for the development of EPS is possible on the basis of "calculated reliability," in the practice of controlling EPS requires the selection of a "standard (directive)" model for the reliability of EPS.

1.6 The Nature and Accuracy of the Initial Information Used

Calculations of indicators of reliability of functioning of the projected EPSs with the help of mathematical models are carried out on the basis of initial information on the prospective operating conditions of the system and the predicted reliability characteristics of EPS elements. Without this information, it is not possible to calculate the reliability of systems.

Since the results of calculations are intended for making decisions on the development of EPS, an important task is to ensure sufficient authenticity and objective value of practical calculations of reliability.

Authenticity depends on the degree of correspondence of used the model to the object and its accuracy, as well as on the completeness and authenticity (accuracy) of the initial data. The inaccuracy of the latter may cast doubt on the expediency of calculating the reliability of complex systems and cause distrust of the obtained numerical values of the indicators even in the presence of a sufficiently accurate calculate model of the system.

Thus, the methodic for accounting for reliability in the design of EPS requires, along with the development of a calculation model and an assessment of the conditions for its use, also an assessment of the properties and the definition of requirements for the initial information.

In real conditions, some information can always be known only approximately. When designing, some of the data are inevitably random or vague. In this case, the properties of the initial information are understood as its stochastic characteristics and their stability (or, conversely, uncertainty) when the external conditions change.

When planning the development of EPS, the problem of uncertainty of the basic initial data on the operating conditions of the projected system in the future, namely the level and mode of power consumption, is topical; provision of necessary fuels; possible results of scientific and technical progress for the planned period.

In addition to the uncertainty of some information, some initial data have a probabilistic nature (ensure HPP water and random fluctuations of load), and some cannot be specified exactly, as, for example, most electrical equipment parameters used in calculations.

In calculating the reliability of EPS, we also have to solve one more information problem—the need to predict the reliability indicators of EPS equipment for the perspective. The basis for this prediction is the reporting data on the operation of systems, statistical data on the operation of their elements. Equipment reliability indicators are obtained as a result of statistical processing (operational or experimental) materials. This treatment allows us to identify the objective trends in the change of these indicators for the future. The situation is more complicated with the evaluation of reliability indicators of new equipment, the statistics of which are not available.

Consider the nature and ways of taking into account the uncertainty, probabilistic nature, and error of the initial information in the design of EPS.

There are two types of uncertainty: the principal (objective) and uncertainty that arose due to the poor organization of the information system. In the past, great importance was attached to the improvement of information systems in the energy sector [46, 47], which cannot be said of the present tense with regard to Russia due to known reasons. Of course, the principal uncertainty will never be eliminated, so the specialists were engaged in developing methods of controlling systems under the conditions of this factor [48–51].

Undefined initial data, as a rule, are set in the form of a range of possible values. At the same time, sometimes the form of data distribution is suggested, most often—uniform distribution. Both the range of possible values and the probability distribution are specified either on the basis of retrospective data analysis [52] or expertly [53].

The undefined information presented in this way requires special organization of studies of variants of possible combinations of initial data ("states of nature") with the use of methods of factor experiment and application of special criteria for the choice of solutions under these conditions [50, 51].

Thus, in the case of indefinite initial information, additional features of the solution arise: The laboriousness of calculations increases, it becomes necessary to introduce new criteria for selecting solutions, and the solutions themselves are multivariate. In the presence of indeterminate initial information, the solution eventually also turns out to be undefined.

The probabilistic nature of the part of the initial data, and especially the parameters of damageability and restorability of the equipment, predetermines the probabilistic and multivariate nature of the problems being solved and the need to determine the numerical characteristics of random processes in place of deterministic ones.

The inaccuracy of the initial data can be of two levels: inaccuracy of the primary information and the inaccuracy of equivalenting in the development of calculated schemes and states (modes). Primary information is always in some way inaccuracy, and the size of the inaccuracy is often unknown in advance and therefore additional researches of this inaccuracy are required.

The inaccuracy of equivalenting, as well as the uncertainty of information, grows with the increase in the lead time for making a decision. This is explained by the fact that the longer the lead time, the less detailed the calculation schemes are developed and the initial factors of the system functioning are presented more enlarged.

In design practice, the estimation of the inaccuracies in the initial data, their influence on the sought parameters, and the accuracy of the various types of equivalenting have been given great attention [52, 54].

Regarding the indicators of reliability of EPS elements as necessary initial information for calculating the reliability of systems, the following recommendations are given in the technical literature. Forecasting of the specified data is carried out by the developed methods of extrapolation of trends for the perspective. For EPS equipment, such tendencies, as a rule, are clearly expressed. This is especially true for trends in the change in reliability indicators of generating block units, depending on their power [55, 56] and power transmission lines from the voltage class [16, 57, 58]. These trends allow us to determine with a good degree of the authenticity indicators of reliability for the perspective both for the exploited and for the new equipment, the statistics on the functioning of which is either insufficient or absent altogether.

In the opinion of the authors, the analysis of the existing spread of indicators reliability of equipment of various EPSs is of current importance. If it is advisable to use indicators averaged over all systems for the long-term perspective, then for the nearest periods it is obviously worthwhile to take the updated (specified, verified) regional indicators if the latter's differences from all-Russian ones are objectively grounded (the difference in climatic conditions, operating modes, etc.).

In this work, we also summarize the results of the researches carried out on the effect of changes in the initial data (factors) on the sought indicators of reliability, and also formulate practical recommendations on the requirements for the initial information and the forms of its presentation in the calculation model, which are the result of accumulated experience in assessing and analyzing the reliability of the main structure of EPS at the level of the UPS, IPS, and RES.

Summarizing the above in this section, we can draw the following conclusion.

Objectively existing information on the operating conditions of EPS in the perspective, in spite of a number of complicating situations, does not render hopeless the solution of the problem of assessing the reliability of EPS in planning their development. But uncertainty or probabilistic nature of individual data, for their account requires a consideration of the increased number of options reflecting various possible conditions for the operation of EPS. In the case of inaccurate information and equivalence, one should focus on obtaining only the quantitative and qualitative characteristics of the system's reliability and, accordingly, the application of the latter for a comparison of options for the development of EPS.

To obtain objective results of solving the reliability problem, preliminary careful work is required to collect, analyze and prepare authenticity initial information, taking into account its undefined or probabilistic nature in a number of cases.

1.7 The Review of Models and Problems of the Analysis of Reliability EPS in a Modern Stage

The previous paragraphs reflect the main provisions and trends in the investigation of the reliability of EPS. However, on the way to their successful realization and introduction into the practice of the control of systems there are significant difficulties, the elimination of which is an urgent task. Below is a brief overview of existing models and programs, and on this basis—a review of problems in the study of reliability of EPS, the main prerequisites and possible directions for their solution.

1.7.1 Brief Review of Existing Models and Programs for Assessing the Reliability of EPS

By the second half of the twentieth century (60–80s), a significant number of models, algorithms, and computational programs were created to solve the problems listed in Sect. 1.4. These programs are used to calculate the reliability indicators of distribution networks [22, 26, 31, 59–62], the power objects of the system (stations, substations and transmission lines) [22, 26, 31, 63, 64], and the basic structure of EPS [32, 33, 65–71]. For the developed models and programs for calculating reliability, the application of various methods described in Sect. 1.5 is characteristic. To investigate the reliability of the basic structure of EPS, the reliability calculation programs for so-called concentrated systems (systems with transfer capabilities of ties that are sufficient to ensure power flows in any possible modes) were often used [11, 68, 72, 73] and systems consisting of from two "concentrated" areas and ties between them [11, 15, 74].

The developed models take into account the majority of significant factors affecting the level of reliability of EPS.

In these models, the accident rate and idle time in planned and capital repairs of generating equipment are quite simply and accurately taken into account; graphs of loads of consumers; the probabilistic nature of the individual initial data (e.g., the load, the water supply of HPP, etc.).

The unification of EPS required the creation of models for estimating reliability for complex configuration systems [34, 71, 75–84], since models for one or two "concentrated" systems connected by a line cannot always be successfully applied in this situation.

The peculiarity of models for assessing the reliability of complex systems in comparison with previous models is the accounting of throughput capacity and the failure rate of electrical ties between the areas of the system, that is, accounting for the so-called "weak" ties [11].

In addition, there arises the need to take into account the correlation dependence between node loads, the problem arises of the optimal distribution of generated power in the regions of the system in order to minimize deficits and undersupply of electricity and a number of serious problems in the computational plan, which will be discussed below.

In the above sources, a description is given of models for estimating the reliability of this class of systems. The number of developed models, of course, is more. Below is a brief description of the models from these publications. In the opinion of the authors, it is these models that are the most developed and the most complete, due to the consideration of a larger number of factors and/or account them into more complete form.

As the EPS developed, new requirements were introduced to the system models. Beginning in the 1970s, a model is developed where a non-concentrated node is considered, and a multi-node system, in which reliability factors are affected not only by the generating equipment, but also by the network component of the basic structure. Such models were created in the leading domestic scientific energy organizations, such as the G.M. Krzhizhanovsky Power Engineering Institute (ENIN) [32], Kirghiz Research Department of Energy (KirNIIEO) [33], Siberian Energy Institute (SEI) [85], etc.

In Russia, the development of existing models continues and the development of new programs for assessing the reliability of EPS. Below are some of these models.

The approach described in Ref. [71], because of the limitations of the configuration of the network under investigation, had a narrow range of applications. The resulting reliability indicators for the nodes are largely determined by the algorithm used in the model for the distribution of power deficits across the nodes of the system. In some cases, this distribution may be far from the real one, determined by the operational strategy of limiting consumers. Thus, with a sufficiently trusting attitude to the calculated system reliability indicators, the practical expediency of the obtained nodal indicators is called into question.

Analyzing the algorithms outlined in [78, 79, 86], we can draw the following conclusions:

1. The use of equivalent transformations of the network scheme allowed us to formulate a sufficiently effective and in a certain sense universal algorithm of the analytical model for estimating the reliability of an EPS of an arbitrary configuration.
2. The approximation of the distribution functions of loading states and of generation by theoretical laws (normal, Charlier or Pearson) somewhat reduces its accuracy and area of application.
3. The algorithm is characterized by the shortcomings that were indicated in the presentation of the paper [71], due to the use of a similar method for the distribution of capacity deficiency (CCD, see Appendix E) between the nodes of the EPS.
4. The scheduled repairs for elements of EPS are not account.

In [81], the author of the "Potok" program for calculating reliability indicators of EPS by the method of statistical modeling asserts that analytical methods of calculation are not always acceptable in analyzing complex EPSs, since these methods "do not fundamentally allow to consider a number of cause-effect phenomena arising in the process functioning of EPS. This is, first of all, the consequences of failures, the development of accidents, the change in the calendar time for scheduled repairs and the like."

The advantage of the model is the use of the chronological principle of modeling (random processes), which allows to take into account the possibilities of the influence of the previous state on the subsequent ones. In the program for calculating the capacity deficit, only the first law of Kirchhoff is taken into account. Two models are used to estimate the minimum deficit in the system. The first uses only balance equations and is based on streaming algorithms [87], and the ambiguity problem is solved with the help of a special problem. The second model is based on the network equations in DC idealization and for its solution the internal points method combined with the relaxation method is applied.

But the developed programs, having the essential advantages mentioned above, have a number of properties that should be attributed to its shortcomings:

(1) a long billing time; to reduce it was required to introduce a number of simplifications and assumptions that significantly affect the accuracy of the calculated reliability indicators;
(2) the results of minimizing the capacity shortage by the Ford–Falkerson method depend on the adopted numbering of the nodes and do not give a real picture of the consequences of failures.

In [75, 84], models of reliability estimation of complex EPS (full and simplified) are presented, based on approximate approximation of stochastic processes in systems, mainly by normal laws and minimization of capacity deficit with the help of approximate calculations of fluxes in ties. The simplified model does not take into account the dependence of damage on the type of restriction and uses the absolute daily maximum instead of the daily load schedule, assuming that during the day the maximum deficit time cannot exceed 0.67 T_{daily} (16 h), as well as a number of other simplifications.

The «Orion» model (the Komi Scientific Center of the Ural Branch of the Russian Academy of Sciences) [88, 89] uses both the dual simplex method and the statistical modeling method. The capacity deficit minimization subroutine takes into account only the first Kirchhoff law, and to estimate the individual state a dual simplex method is used, in which the initial approximation for each next state is taken to be the results of the calculation of the previous state. The advantage is that the definition of the initial plan in a dual task is carried out without difficulty.

Abroad, just as in our country, models for assessing the reliability of EPS have developed. From the first, foreign models can be identified the following. Model PERU (France) [77], in which the evaluation of a complex system was replaced by a sequential analysis of two-node scheme. On the one hand, such a method is imperfect, on the other, even such a model can help in making any responsible decisions on the development of EPS. There is also the MEFISTO model (Belgium) [82]; Model A and Model B (USA) [76].

Of the later developed foreign models for assessing the reliability of the basic structure of EPS, the MEXICO model (France) can be singled out [90, 91]. This model appeared in the early 80s and continues its development. It is based on statistical modeling and when modeling in time does not take chronology into account. A particular state of the system is modeled randomly in accordance with the availability factors of the generating and transmitting equipment. For each such state at a given level of consumption generating capacities and streams in lines are determined solving of task by linear programming, which minimize the amount of running costs and damage from undersupplying electricity when accounting for network equations in DC idealization. Then the results of all the considered states are averaged, and various reliability indicators are calculated. The SICRET model (Italy) [92] is similar to MEXICO, but has a number of differences. It allows you to consider AC and DC networks. The COMREL model (Canada) [93] uses the analytical method. For the analysis of each state, one of three approaches can be used: a linear flow model, a DC idealization model, and a fast, separated flow distribution method. There is also a model TRELSS (Poland) [94] and model CONFGT (Brazil) [95, 96].

Currently [97] system operators in the USA use the GE MARS program, some—the Grid View. The power company PJM relies on its own software resources developed within the company (PRISM). New Zealand uses the GEM. In the past, the software complex MARELI was used. Australia uses both its own set of programs and relies on the involvement of consultants. The California Energy Regulatory Commission uses its own SAM (Supply Adequacy Model) model. In the literature, there is a mention of the use by the System Operator of California of the NARP and PLEXOS software complexes for calculation of balance reliability.

GE MARS, Grid View, PLEXOS, and NARP have chronological Monte Carlo simulation algorithms. The MARELI model is based on the convolution algorithms of distribution laws.

- GE MARS is the most widely used complex. At the same time, it has separate algorithmic drawbacks; the most significant is the transport model of the network.
- The Grid View software package was developed by the same team of analysts who developed GE MARS and (according to the information received from its developers) possesses all the functional properties of GE MARS, and also offers a number of additional features, including calculation of the full electrical network. The main functional purpose of Grid View is to perform calculations for the optimal selection of the composition of generating equipment and optimal modes. Balance sheet reliability calculations as a function were added later. It is built on modern programming techniques and unlike GE MARS offers a graphical user interface.
- MARELI does not propose a chronological methodology for modeling using the Monte Carlo method and has significant limitations on the configurations of power transmission systems that it can simulate.
- NARP has interesting algorithmic capabilities that surpass those of GE MARS, but, on the other hand, little is known about the practical application of this software.
- PLEXOS, like Grid View, is a widely used software package for modeling the choice of the composition of the included generating equipment and modes. The possibility of calculating balance reliability by the Monte Carlo method seems to have been added recently.

1.7.2 The Main Problems of the Reliability Analysis of EPS

From the review of existing models of reliability assessment of complex EPSs, given in Sect. 1.7.1, it can be concluded that these models and corresponding programs have shortcomings, including those that hinder their wide implementation in the control of EPSs to date.

In this regard, the task of creating models and computational programs that are devoid of some shortcomings continues to be relevant. At the same time on the way of creating effective programs for calculating reliability indicators of the main structures of complex modern EPS, there are the following main problems and difficulties.

1. **Estimating the transfer capability of ties** in a complex system and more accurate accounting for them in a reliable model. The necessity of this is caused by one of the main features of EPS and complicates the calculation of their reliability. It is known that transfer capability of ties, determined by the stability of their work, are functions of the composition of the operating equipment and the system mode, while the values of the transfer capability limits of the links in individual cases can vary quite widely.

It should be noted, however, that this problem in Russia is relevant today for the main domestic networks at 220 kV and above. At a voltage lower than 220 kV and for a number of ties of a higher voltage level, the throughput is determined by the thermal mode of power transmission and, consequently, the difficulty of estimating the maximum transmitted power through such ties is reduced.

2. **Accounting for the specifics of the operation of HPPs, CHPPs, and PSPs**. The developed models for assessing the reliability of complex EPS either do not take into account at all, or take into account very closely such features of the operation of these objects as seasonal and long-term limitations of hydroelectric power stations on the energy of the watercourse; dependence of the use of the available capacity of CHPP on thermal loads; the possibility of using the available capacity of the PSP according to the conditions for filling the reservoir.

At present, there are no sufficiently complete and reliable results of studies of conditions in which it is necessary to take into account the peculiarities of the operation modes of HPPs, CHPPs, and PSPs. It is obvious only that the significance of these factors is determined by their share in the system under study, and also by the degree of manifestation of these factors. In general, the issue of the need for such accounting remains open.

3. **Accounting for the reliability of the supply of EPS power plants of primary energy resources**. The importance of accounting for this factor (the daily supply on hydroelectric power plants water and TPP fuel) is undoubted and becomes especially important in the situation when there are difficulties in supplying fuel, interruptions in its delivery to the station due to unreliable operation of transport systems. In existing models, especially foreign ones, this circumstance has so far

been of little importance, since the provision of energy resources was considered absolute.

4. **Minimization of capacity in calculated states (modes)**. This task is one of the most important and most difficult in the calculation of the reliability of complex EPS, since the magnitude of the capacity deficit in them depends not only on the total unbalance of capacity, but also on the transfer capacity of the ties (which, as already noted, themselves in some cases are functions of the regime), as well as from a decrease in the quality of electricity from the consumer below the permissible level. In addition, the conditions for minimizing the deficit are determined by the economic characteristics of consumers—the values of the damages from the interruption of electricity supply, or their categorization, as well as the dispatcher policy of load limitation. All this assumes minimization of deficits based on electrical calculations of EPS states (regimes).

Calculations of modes of complex EPS are a time-consuming computational operation, even more complicated when trying to optimize them in conditions of power limitations.

At present, there are virtually no developed algorithms and programs for minimizing capacity deficient (or damage from power supply limitations) based on calculations of the corresponding electrical regimes. There are only algorithms and programs that use methods of mathematical programming for this purpose, in particular the methods of minimum cross sections and of maximum flows, the drawbacks of which, as mentioned above, not accounting of specificity of the EPS modes and the ambiguity of the solution. For example, the application of the effective Ford–Falkerson algorithm [87], as already indicated, leads to the dependence of the calculated values of capacity deficits in the calculated nodes of the system on the order of their numbering in the calculation scheme.

This circumstance is unimportant in determining the reliability index of the system as a whole, but it makes it impossible to assess the reliability of the calculated nodes (subsystems) entering the system. However, the evaluation of the reliability of subsystems is the basis for solving many practical problems of control of EPS.

On the other hand, in a number of cases the need for calculating the electric regime is eliminated because of the deep equivalence of the design scheme of the system under research. Electric calculation of the regime is impossible, if each calculated node in the scheme displays a set of several substations of different stress levels and connections between them (of internal ties of the concentrated node). In this case, it is sufficient to calculate the regime in a simplified manner, for example, only taking into account Kirchhoff's first law.

5. **The problem of an extremely large number of likely states of EPS**, which, as a rule, is characterized by a complex system and which, when calculating reliability, should be evaluated in terms of the scarcity of electricity supply. For complex calculation schemes, the number of possible states is so great that the estimation of all states even with the help of modern computer technology is impossible. Moreover, this is unreal if it comes to optimizing the states taking

into account the specificity of their electrical regimes. In the existing algorithms for calculating the probable states of systems, the following outputs are used:

(1) We consider not G_1 states of the EPS for equipment ($G_1 = 2^I$, where I is the number of elements), and G_2 of states by power. In this case $G_2' = P_{\text{avail}}^{\text{g}} / P_{\text{step }m}$—for generating equipment and $G_2'' = P_n / P_{\text{step }n}$—for electrical ties [98, 99, etc.]. In these formulas, $P_{\text{avail}}^{\text{g}}$ is the available generating power in the calculation node, P_n is the total ties capacity.

The number of calculated states decreases with increasing power step P_{step}, but an increase in P_{step} leads to an increase in the calculation error, and on the other hand, in the case of a system of M nodes and N connections between them, the number of calculated states increases with increasing M and N:

$$G_{\Sigma} = \prod_{m=1}^{M} G'_{2m} \prod_{n=1}^{N} G''_{2n};$$

(2) not all states are considered, but the most probable ones [98, 99]. In this case, the complete series of the distribution of the states of the corresponding equipment is limited to the number of terms $G_{2\,\text{lim}} < G_2$ whose probabilities exceed the selected value;

(3) a frequently used technique is to consider only those states that are characterized by simultaneous idle time in post-fault states of not more than one, two, or (very rarely) three elements of the system. This method, suitable for systems with a small number of highly reliable elements (no more than 20–30), leads to significant errors in the analysis of the reliability of modern systems, the number of elements in which can sometimes be more than a few of hundred or thousands.

6. **Accounting for the compatibility of loads of areas of the system located in different time zones**. This question becomes relevant in the research of geographically extended systems, when it becomes necessary to take into account the time shifts in individual regions.

In addition, when analyzing reliability for practical purposes, it is important to identify deficiency states (regimes) associated not only with the periods of maximum loads (as is done in most developed programs) but also the deficiency of certain special states (periods of minimum loads, seasonal regimes, etc.).

All this makes it impossible to use the annual load schedules for the duration and requires consideration of the hourly load schedules taking into account the waist shifts [100].

7. **The problem of harmonizing the results of solving reliability problems at different territorial levels and time stages**. The global task of assessing the reliability of EPS in accordance with the principles outlined in Sects. 1.3 and 1.4 is divided into a number of separately solvable tasks for the relevant technological links of the system, territorial and temporal levels. At the same time, of course,

there is a problem of interrelating the obtained results of reliability calculations of different levels, stages, and links.

Specially relevant is the coordination of reliability assessments at the territorial levels and technological links of the system.

When calculating the reliability of the basic structure of EPS, the existing techniques are oriented toward determining the reliability indexes of equivalent loads of the calculated schemes of the EPS of the corresponding level (UPS, IPS, or RES). The norm of reliability of power supply for consumers for the specified calculations in [15] is given without indicating to what level it relates and what is the resultant reliability of the power supply of consumers corresponds.

However, such indicators are sufficient to choose the variant of the development of a particular level or link of the EPS by comparison. To estimate the resulting levels of reliability of power supply to groups or individual consumers, depending on the reliability of both the distribution network and the basic structures of EPS at different levels (UPS, IPS, RES), in this case is not possible.

To assess the reliability of specific load nodes or individual consumers, an appropriate calculation of the reliability of the area of the system is required, taking into account the conditions of its operation within the corresponding RES, and the RES—within the framework of IPS, and the latter within the UPS, and for these conditions, appropriate quantitative indicators and methods their definitions.

It should be noted that, despite the paramount importance of this issue, it is given insufficient attention. There are no specific proposals in this regard.

8. **The choice of factors, which are necessary for the analysis of reliability of EPS**. In studies of reliability, much attention is paid to the analysis of the impact of the conditions (factors) of functioning on the failure-free operation of electricity supply to consumers. At the same time, the created models for assessing the reliability of EPSs serve as a tool for carrying out this analysis. The results of research (quantitative relationships) find application both in engineering practice when making appropriate decisions, and for improving models.

Not only factors as such, but also generalized characteristics of factors and various forms of their representation in reliability models are exposed to the research. As a result, the degree of importance of each factor is revealed and recommendations are given on the most rational ways to take them into account in algorithms and programs. There is a large number of works in which the influence of various factors is analyzed.

So in Lyalik [101], in particular, the results of the research of the effect of the load schedule on the level of reliability of the system are presented. A number of investigations have established a relationship of reliability with a degree of redundancy [67, 101], the supply of HPPs [102], random load fluctuations, etc.

However, sometimes different researchers make different (sometimes directly opposite) conclusions from their analysis of the factors affecting the reliability of EPS (for more details, see Chap. 4).

In this situation, apparently, the work on the analysis of factors should continue to eliminate contradictions, as well as generalization of known results to a greater number of possible cases.

9. **Evaluation of the effect of system elements on the level of reliability**. An important problem of researching the reliability of systems is the ability to assess the "responsibility" of each element or link of the system for its reliability or reliability of individual subsystems. Solving this problem would allow us to obtain constructive calculation results that would become the guiding material for the application of measures to effectively increase the reliability of EPS.

* * *

The main task of this research, taking into account the modern state of the issue outlined in this chapter, is the development of a means for investigating the reliability of the basic structures of complex EPS (system reliability).

On the basis of this methodic, the possibility of creating methodical provisions for the calculation of reliability in the design of EPS is explored, which would make it possible to use the results obtained in this work in practice. In this regard, attention is paid to the analysis of the existing system of accounting for the reliability of EPS in the design practice and determines the location of the proposed methodic and model of reliability assessment among other calculate methods of design.

The basis of the methodology should be the model created for this purpose and the corresponding computational program for determining the reliability indexes of the EPS of a complex configuration. Algorithm and program, of course, should be guided by the latest modifications of modern computer technology.

Using the most successful solutions of individual issues in the existing models for assessing the reliability of both "concentrated" EPSs and EPS with "weak" ties, in the model being developed, so that it can meet the set goal, special attention, as follows from Sect. 1.7.2, should be given:

- accounting for the throughput of ties and the possible mutual support of subsystems in scarce states;
- solving the problem of minimizing capacity deficit in the calculated states of the system, taking into account the specifics of the states (regimes) of complex systems and the strategy of limiting consumers adopted in practice;
- determination of quantitative indicators that would ensure the mutual consistency of reliability calculations at various hierarchical levels of EPS;
- ensuring the performance of mass calculations, for which the problems of reducing the dimensionality of the set of calculated states of the system should be solved and attention paid to some other points of a methodical and computational nature, including the parallelization of the account and other modern computing and information technologies.

The task of analyzing a number of factors influencing the reliability of EPS is also posed. Evaluation of the influence of factors was carried out by calculating the reliability of some real structures of modern EPS on the created computational model. On the basis of the researches carried out, recommendations are given on the permissible accuracy and methods of taking into account the influencing factors in the computational models for estimating the reliability of EPS. In addition, the task is to determine ways to effectively increase reliability based on the calculation of pose reliability indicators of EPS for the purpose of justifying the structure of systems.

Finally, in the work, on the one hand, the task is to check the developed methodic for calculating reliability, algorithm and computational program by analyzing some variants of the prospective structure of the EPS, and on the other hand, the task of developing specific recommendations on the reliability of the EPS for a period of up to 15–20 years based on the use of the proposed methodic.

References

1. Ushakov IA (1967) Basic principles and methods of the theory of reliability. Voprosy filosofii 6:103–121 (in Russia)
2. Ferman RA (1932) A method for the economic evaluation of the operational reserve of electrical systems, Elektrichestvo 20:944–946 (in Russia)
3. Kukel-Kraevsky SA (1935) On the minimum required reserves in steam turbo-generators for the electric power systems of the USSR. Elektrichestvo 17:26–29 (in Russia)
4. Malaya RY (1935) On the problem of reserve in power systems. In: Material of the first all-union meeting on the operation of power supply systems, no 2. ONTI, Moscow-Leningrad, pp 389–392 (in Russia)
5. Tourover JM (1935) The main issues of reliability of electrical systems. In: Materials of the first all-union meeting on the operation of energy-supply systems, vol 2. ONTI, Moscow-Leningrad, pp 393–416 (in Russia)
6. Afonin NS (1934) Damage from the shutdown of consumers. Electr Stat 7 (in Russia)
7. Yakub BI (1934) Indicators and methods for calculating reliability in the energy sector. Electricity 18. S. 1 13 (Discussion in the same place: 1935. № 8, pp 47–48) (in Russia)
8. Smith SA (1934) Service reliability measured by probabilistic of outage. Electr World 103:371–374
9. Smith SA (1934) Spare capacity fixed by probabilistic of outage. Electr World 103:222–223
10. Gnedenko VV, Belyaev YK, Solovyov AL (1965) Mathematical methods in the theory of reliability. Nauka, Moscow, 524 p (in Russia)
11. Rudenko YN, Cheltsov MB (1974) Reliability and redundancy in electro-energy systems. Methods of research. Science, Siberian Branch, Novosibirsk, 264 p (in Russia)
12. Reliability of energy systems (a collection of recommended terms). (2007) IAC "Energia", Moscow, 192 p (in Russia)
13. Buslenko NG, Kalashnikov VV, Kovalenko IN (1973) Lectures on the theory of complex systems. Soviet radio, Moscow, 440 p (in Russia)
14. GOST 27.002-89 (1990) Reliability in technology. Basic concepts. Terms and definitions. Publishing house of state standards, Moscow, 37 p (in Russia)
15. Methodical instructions for designing the development of power systems (2003). Izd-vo NTs ENAS, Moscow (in Russia)
16. Rokotyan SS, Shapiro IM (eds) (1985) Handbook on the design of electric power systems. Energoatomizdat, Moscow, 352 p (in Russia)
17. Malkin PA, Kovalev GF (1981) Software for the calculation of reliability in the design of EPSs. In: Methods for investigating the reliability of large energy systems. issue 22. SEI SB AS USSR, Irkutsk, pp 7–16 (in Russia)

18. Methods and Techniques for Reliability Assessment of Interconnected Systems (1997). Prepared by CIGRE T.F. 38.03. 11. July 1997, 60 p
19. Billinton R, Allan R (1988) Evaluation of the reliability of electric power systems: translation from English. Energoatomizdat, Moscow, 288 p (in Russia)
20. Andrenyi J (1983) Simulation in the calculation of reliability in electric power systems: translation from English. In: Rudenko YN (ed). Energoatomizdat, Moscow, 180 p (in Russia)
21. Kovalev GF, Lebedeva LM (2015) Reliability of electric power systems. Publisher "Science", Novosibirsk, 224 p (in Russia)
22. Rozanov MN (1984) Reliability of electric power systems. Energoatomizdat, Moscow, 200 p (in Russia)
23. Guk YB, Losev EA, Myasnikov AV (1974) Evaluation of the reliability of electrical installations . In: Konstantinova. Energia, Moscow, 200 p (in Russia)
24. Mostkov MA (1960) On the emergency reserve of power and energy in the electric power system. Izvestiya AN SSSR. Department of Technical Sciences. № 6, pp 828–836 (in Russia)
25. Guk YB, Kazak NA, Myasnikov AV (1970) Theory and calculation of reliability of power supply systems. In: Fedosenko RY (ed). Energia, Moscow, 117 p (in Russia)
26. Sinchugov FI (1971) Rationing of reliability in power systems. Power plants 10:8–12. The discussion is in the same place, 1972, no 5, pp 81–82; 1972, no. 8, pp 88–89 (in Russia)
27. Manov NA (1973) Probabilistic analysis of the reliability of the link of the generation of the electric energy system with the passage of the annual maximum loads. In: Proceedings of the Kirghiz research department of energy, vol 4. Publishing house "Kyrgyzstan", Frunze, pp 28–34 (in Russia)
28. Kozlov BA, Ushakov IA (1975) Handbook on calculating the reliability of radioelectronics and automation equipment. Soviet Radio, Moscow, 472 p (in Russia)
29. Rozanov MN (1967) Technical and economic comparison of electrical supply systems that provide different degrees of reliability. In: Reliability of power supply. Sat. articles. Energy, Moscow–Leningrad, pp 9–20 (in Russia)
30. Bolotov VV, Artyugina MM, Burtseva GE, Dolgov PP (1970) Theoretical questions and methods for designing energy systems. Science, Leningrad, 274 p (in Russia)
31. Sinchugov FI (1971) Calculation of the reliability of electrical connections schemes. Energia, Moscow, 176 p (in Russia)
32. Volkov GA (1968) Determination of the optimal reserve of active power in the unification of power systems, Izvestiya AN SSSR. Power Transp 4:79–82 (in Russia)
33. Itkin EA, Kolosov IS (1970) To the development of the technique for allocating a capacity reserve in the integration of power systems at the stage of planning their development. In: Problems of energy and use of energy resources in Kyrgyzstan. Ilyam, Frunze, pp 71–84 (in Russia)
34. Itkin EA, Kolosov IS (1970) The choice of bandwidth interconnection in the association with allowance for redundancy in individual systems . In: Reliability and profitability of power systems (vol I). Reports of scientific and technical conference on the generalization of experience in the design, construction and operation of power transmission lines and substations in the regions of Siberia and the Far East. Science, Sib. Branch, Novosibirsk, pp 109–114 (in Russia)
35. Itkin EA, Kolosov IS (1970) Placement of the reserve in the integration of energy systems of complex configuration. In: Theses of the reports of the republican seminar "Methods for calculating the emergency reserve in complex power systems". Ilyam, Frunze, pp 17–22 (in Russia)
36. Sobol IM (1968) The Monte Carlo method. The main edition of physics and mathematics. Izd-vo Nauka, 68 p (in Russia)
37. Kulesha R (1968) Reliability problems of large systems. In: Optimization and control methods in large systems. Ed. Bolg. AN, Sofia, pp 137–168
38. Ushakov IA (1974) Construction of highly reliable systems. Knowledge 62 (in Russia)
39. Oboskalov VP (2002) Reliability of providing power balance of electric power systems. GOU VPO USTU–UPI, Ekaterinburg, pp 210 p (in Russia)

40. Rozanov MN (2000) Reliability of electric power systems, vol 2. Energoatomizdat, Moscow, 568 p (in Russia)
41. Kovalev GF, Lebedeva LM (2000) Model of the reliability assessment of electric power systems in the long-term planning of their work. Electricity 11:17–24 (in Russia)
42. Fokin YA (1985) Probabilistic–statistical methods in the calculation of power supply systems. Energoatomizdat, Moscow, 210 p (in Russia)
43. CIGRE (1987/1992) Power system reliability analysis—application guide, Paris, 1987; Power system reliability analysis—composite power system reliability evaluation, Paris, 1992
44. Wentzel ES (1972) Investigation of operations. Sov. Radio, Moscow, 561 p (in Russia)
45. Aleksandrov IA, Kovalev GF, Rudenko YN (1973) Comparative analysis of reliability of various options for the development of electric power systems. In: Reports at the third all-union scientific and technical meeting on the stability and reliability of the energy systems of the USSR. Energia, Leningrad, pp 394–405 (in Russia)
46. Kudryashov GR, Rudenko YN, Cheltsov MB (1967) Problems of information collection and processing for analysis of reliability of electric power systems operation. In: Materials to the all–union conference on energy efficiency for account of optimization of power supply systems and power consumption regimes of non-ferrous metallurgy enterprises. KPI, Krasnoyarsk, pp 73–88 (in Russia)
47. Gladyshev GP, Kaplun SM, Kudryashov GR et al (1974) Problems of developing a system of information on the reliability of the equipment of the electric power industry, Izvestiya AN SSSR. Energy Transp 6:10–18 (in Russia)
48. Wagner G (1973) Fundamentals of operations research, vol I. Mir, Moscow, 336 p (in Russia)
49. Yudin DB (1974) Mathematical methods of control under incomplete information. Soviet Radio, Moscow, 400 p (in Russia)
50. Belyaev LS, Voitsekhovskaya GV, Zeyliger AP, Khabychev YL (1973) Properties of the initial information used in optimizing the development of electric power systems, and the variety of its formalized representation. In: Accounting uncertainty of the initial information, issue 1. SEI SB AN SCCR, Irkutsk, pp 14–24 (in Russia)
51. Belyaev LS, Makarov AA (1973) A general way of solving problems in conditions of non–definiteness. In: Reckoning the uncertainty of the initial information, issue 1. SEI SB AS USSR, Irkutsk, pp 25–35 (in Russia)
52. Golinets MV, Zeyliger AN, Khabachev LP et al (1970) Analysis of the errors of the initial information and their influence on solutions adopted in the perspective design of power systems. In: Optimization and control in large energy systems. Tom. 3. SEI SB AS USSR, Irkutsk, pp 229–251 (in Russia)
53. Guk YB, Okorokov VR, Shapot DV, Shchavelev DS (1974) Complex optimization of technical solutions in power engineering under conditions of uncertainty. In: Factor of uncertainty in making optimal decisions in large energy systems, vol 1. SEI SB AS USSR, Irkutsk, pp 118–124 (in Russia)
54. Voropai NI, Gamm AZ, Krumm LA et al (1970) On the problem of equivalence in the construction of mathematical models. In: Optimization and control in large energy systems. Tom. 1. SEI SB AS USSR, Irkutsk, pp 193–218 (in Russia)
55. Kudryashov GR, Rudenko YN, Fedotova GA (1970) Reliability indicators of boiler and turbine units of the electric power systems of Siberia. In: Reliability and profitability of power systems, vol 1. Reports of the scientific and technical conference on summarizing the experience of design, construction and operation of power transmission lines and substations in the regions of Siberia and the Far East. Science. Sib. Ltd., Novosibirsk, pp 181–198 (in Russia)
56. Gladyshev GP, Gorin VP, Yakovlev GG (1975) Reliability of blocks of thermal power stations of the USA in 1960–1071. Energy Manag Abroad 2:1–11 (in Russia)
57. Melnikov NA (1969) Electrical networks and systems. Energia, Moscow, 496 p (in Russia)
58. Rokotyan SS Operational characteristics of ultrahigh voltage lines. Energy Econ Abroad 6:23–27 (in Russia)

59. Kholmsky AV et al (1967) On the calculation of the optimal reliability of power supply schemes with a minimum of reduced costs, taking into account additional criteria, Izvestiya VUZov. Power Eng 10:17–28 (in Russia)
60. Guidelines for assessing the reliability of electrical networks when choosing the best schemes (1968). CAO VGPI and NII Energosetproekt, Tashkent
61. Fokin YA, Chan DL (1973) Structural analysis and methods for assessing the reliability of complex power supply schemes. Electricity 5:16–24 (in Russia)
62. Rozanov MN (1978) A review of existing methods for calculating the reliability of electric networks. In: Problems of reliability of power systems. Proceedings of VNIIE, issue 55. Energia, Moscow, pp 38–55 (in Russia)
63. Sinchugov FI (1967) Choice of the main circuits of electric connections of block power stations. In: Electric stations, № 5, pp 9–30 (in Russia)
64. Sinchugov FI (1969) Probabilistic method for calculating the reliability of electrical circuit schemes. In: Electric stations, № 12, pp 38–41 (in Russia)
65. Volkov GA (1963) Choice of the optimal value of an incorporated power reserve in the power system, Izvestiya AN SSR. Energy Transp 3:257–265 (in Russia)
66. Rozanov MN (1967) Choice of the optimal degree of redundancy in the design of power supply systems. Electricity 4:25–28 (in Russia)
67. Rozanov MN (1987) Reservation in power systems. In: Reliability of power supply. Sat. articles. Energy, Moscow–Leningrad, pp 216–271 (in Russia)
68. Kryukov AA, Lyalik GN (1970) Simplified methodology for determining the optimum value of the emergency reserve in power systems. In: Problems of the optimal development of power systems and new technical means of their protection. Nauka, Moscow, pp 31–42 (in Russia)
69. Lyalik GN, Urvantsev NP (1972) Determination of the emergency power reserve based on universal characteristics of the specific reserve, № 1. pp 26–30 (in Russia)
70. Kolosov IS, Itkin VA, Polyakova LM et al (1973) Optimization of the reserve of active power in the unification of power systems. In: Reports at the third all–union scientific and technical meeting on the stability and reliability of the energy system of the USSR. Energia, Leningrad, pp 445–454 (in Russia)
71. Volkov GA (1973) Determination of the reliability of power supply in integrated power systems using analytical methods of calculation. In: Proceedings of the III all–union scientific and technical meeting on the stability and reliability of the power systems of the USSR. Energy, Leningrad, pp 518–527 (in Russia)
72. Markovich IM (1969) Modes of energy systems. Energia, Moscow, 352 p (in Russia)
73. Volkov GA (1970) Calculation of the optimal reserve of active power in designing the development of power systems on the Ural–2 computer. In: Problems of the optimal development of power systems and new technical means of protecting them. Nauka, Moscow, pp 3–12 (in Russia)
74. Kryukov AA (1970) Choice of the capacity of interconnections. In: Problems of the optimal development of power systems and new technical means for their protection. Nauka, Moscow, pp 24–30 (in Russia)
75. Nepomnyashchy VA (1975) Mathematical model of the reliability of the electric power system. In: Methodological issues of reliability research of large energy systems, issue 8. Reliability in the design and operation of energy systems. SEI SB AS USSR, Irkutsk, pp 110–118 (in Russia)
76. Casazza JA, Mallard SA (1972) Generation and transmission reliability. Message № 32–11 at the session of the CIGRE 1970 Translation of Golovitseva VI. In: Venikov VA (ed) Regimes of the work of power systems. Energia, Moscow, pp 85–95 (in Russia)
77. Auge J, Bergougnoux J, Dodu JC, Pouget J (1972) Probabilistic study of a transmission system interconnection (PERU Model). Report No. 33–16 at the 1970 CIGRE session. Translated by ISLyseev. In: Venikova VA (ed) Regimes for the operation of power systems. Energia, Moscow, pp 95–104 (in Russia)
78. Vasin VP, Prokhorenko VI (1980) On the construction of the probability distribution function of the total power of the power system // Izvestiya AN SSSR. Energy and transport ». № 6. Pp. 61–67. (In Russia)

79. Itkin EA, Shadrin VA (1975) Model of reliability evaluation of the planned electric power system . In: Methodological issues of investigation of reliability of large energy systems, issue 8. Reliability in the design and operation of energy systems. SEI SB AS USSR, Irkutsk, pp 103–109 (in Russia)
80. Itkin YA, Shadrin VA (1975) Construction of a model for analysis of the reliability of a complex electric power system using approximating curves for describing random variables. In: Methodological issues of reliability research for large energy systems, issue 12. SEI SB AS USSR, Irkutsk, pp 44–49 (in Russia)
81. Mogirev VV (1975) Algorithms and programs for calculating reliability indicators of electric power systems by the method of statistical modeling (program "Potok"). In: Methodological issues of investigation of reliability of large energy systems, issue 4. Mathematical methods of analysis of reliability. SEI SB AS USSR, Irkutsk, pp 24–35 (in Russia)
82. Boleriaux H, Brancart D, Vankelecom J (1976) An original method for computing shortfall in power systems (MEFISO method). Report No. 33 – 09 at the 1974 CIGRE session. Translation from English Tufanov VA. In: Veni–kov VA (ed) Planning the development and operation of energy systems. Energia, Moscow, pp 53–62 (in Russia)
83. Nepomnyashchy VA (1976) Economic–mathematical model of the electric power system for calculating its reliability // In: Methodological issues of reliability research for large energy systems. Issue. 12. – Irkutsk: SEI SB AS USSR, pp. 34–43. (In Russia)
84. Nepomnyashchiy VA (1978) A calculation of reliability in the design of power systems. Energia, Moscow, 200 p (in Russia)
85. Kovalev GF, Rudenko YN, Cheltsov MB (1977) On models for estimating the reliability of electric power systems. In: Modeling of electro-energy systems. Abstracts of the VII all–union conference. TPI, Tallinn, pp 177–178 (in Russia)
86. Itkin EA, Kolosov IS (1973) Estimation of the reliability of complex integrated power systems at the stage of their design. In: Methods for calculating the emergency reserve in complex power systems. Proceedings of the Kirghiz research department of energy, issue 4. Publishing house "Kyrgyzstan", Frunze, pp 21–27 (in Russia)
87. Ford LR, Falkerson DR (1966) Flows in networks. Mir, Moscow, 236 p (in Russia)
88. Chukreev YY, Manov NA (1991) Investigation of the reliability of the basic structure of the electric power system using the Orion software–computing complex. Methodological issues of reliability research of large energy systems, issue 41. SEI of the SB RAS, Irkutsk, 29–40 (in Russia)
89. Chukreev YY, Manov NA, Slobodyan Y V (1987) Investigation of reliability in managing the development of multinode power systems. New Sci Methods 24:27 (in Russia)
90. Dodu AM (1979) Recent improvement of the Mexico model for probabilistic planning studies. IPC Bus Press Electr Power Energy Syst 1:147–153
91. Composite power system reliability analysis application to the New Bruns–wick Power Corporation System. The draft report of the CIGRE Symposium on Electric Power System Reliability, 16–18 Sept 1991, Montreal, Canada
92. Bertoldi O, Scalcino S, Salvaderi L (1991) Adequacy evaluation: an application of ENEL's SICRET program to new Brunswick Power System. In: CIGRE symposium "electric power system reliability", Montreal. WG 38.03/01
93. Salvaderi L, Bilinton R (1985) A comparison between two fundamentally different approaches to a composite system of reliability. IEEE Trans Pas 104
94. Methodologies and tools for electric power system reliability assessment on HL I and HL II Levels, Józef Paska Institute of Electrical Power Engineering Warsaw University of Technology Warsaw, Poland. In: 9th international conference electrical power quality and utilization, Barselona, 9–11 Oct 2007
95. Leite da Silva AM, Mello JCO (1991) Improvements in the composite generation and transmission reliability evaluation. CIGRE symposium "electric power system reliability", Montreal, WG 38.03/01
96. Paska J, Bargiel J, Goc W, Momot A, Nowakowska E, Sowa P (2003) Reliability assessment in electric power system planning and management. In: 7th international conference "electrical power quality and utilization—EPQU'03". Cracow–Poland, 17–19 Sept 2003

97. Lazebnik A, Rudkevich A, Sorokin I (2010) A market–based approach to power systems expansion planning. In Presented at the federal energy regulatory commission technical conference on planning models and software. Washington, DC, 9–10 June 2010
98. Kovalev GF, Olenkevich VP, Patseva TV (1981) Influence of factors of EES function and methods of their representation in models on the calculated value of the reserve of generating capacity. In: Methodological issues of reliability research of large energy systems, issue 23. SEI SB AS USSR, Irkutsk, pp 54–65 (in Russia)
99. Mogirev VV, Rudenko YN (1973) Algorithms for investigating the reliability of electric power systems of arbitrary configuration, Izvestiya AN SSSR. Energy Transp 1:38–45 (in Russia)
100. Lyalik GN, Galaktionov MV, Urvantsev VI (1974) Analysis of the factors that are accounted for in the design determination of the need for emergency power reserve. In: Questions of the rationale and choice of parameters for developing power systems. Proceedings of the Institute "Energosetproject", no. 4. ESP, Moscow, pp 15–34 (in Russia)
101. Lyalik GN (1975) The role and importance of hydroelectric power stations in ensuring the optimal level of reliability of power supply for mixed energy systems and methods for their accounting in the design. In: Reports of the scientific and technical conference on the results of scientific research for 1963–1969 years. Electric power section. Subsection of hydropower. MEI, Moscow, pp 21–33 (in Russia)
102. Manov NA, Slabodyan YV, Kuznetsova IF (1980) Estimation of the reliability of the electric power system (taking into account the peculiarities of the electric power consumption regime). Academy of Sciences of the USSR. Komi Branch, Syktyvkar, 63 p (in Russia)

[illegible]

[illegible]

[illegible]

[illegible]

[illegible]

[illegible]

Chapter 2
Methodic for Research of Reliability of Complex Electric Power System in Their Designing

2.1 Comprehensive Assessment of the Reliability of Variants for the Development of EPS as a Proposed Methodic to a More Accurate Account of Reliability in the Design of Their Basic Structure

The indicated *problems* and analysis of the models (Chap. 1, clause 1.7) for reliability consideration in the practice of EPS designing require scientifically based solutions and their implementation in the models for application to EPS designing.

Needless to say that the problems enumerated in Sect. 1.4 must be solved comprehensively, whereas part of them can be solved independently as before only with coordination of the input and output information with other problems.

The considered reliability problem of *the EPS main structure*, which comprises composition and parameters of generating capacities, configuration, and parameters of a backbone network is closely related with optimization of the size, structure, and allocation of generating capacity reserves and optimization of additional transfer capabilities of ties from the list of problems presented in Sect. 1.4.

Optimization of the volume, structure, and allocation of generating capacity reserves as an important part of the problem of choosing the size, structure, allocation, and commissioning time of EPS generating capacities is closely related with the EPS reliability problem at the system level. The goal of solving this problem is to determine a calculated of the total capacity reserve as the difference between the available capacity and peak load, which consists of the calculated values of the strategic (former national) reserve, reserve for scheduled (current, medium and major) repairs, overloading reserves and emergency reserves; to choose composition and allocation of the corresponding capacity reserves.

In the present-day practice of designing, the required calculations are made based on the approaches developed for this purpose. However, even publications [1, 2] indicated that these approaches were approximate, since they contained some essential assumptions, which decreased validity of the results obtained in some cases and narrowed an application sphere of these methods.

G. F. Kovalev and L. M. Lebedeva, *Reliability of Power Systems*, Power Systems, https://doi.org/10.1007/978-3-030-18736-1_2

Optimization of additional transfer capabilities of ties in a backbone network as part of the problem on choosing the structure and parameters of the EPS network is also a central problem of EPS reliability at the system level. The goal of solving this problem is to adjust the chosen structure and parameters of the EPS backbone network from point of view support of reliability of system operation at deviations from normal operation of EPS generating and network equipment.

At present, the solution of this problem does not have a rigorous justification, and it is based on the accumulated experience the design of systemic ties, empirical relationships and the expert formation of variants for the development of networks, that also, like in the previous problem, reduces the authenticity of the results obtained [3, 4].

However, the disadvantage of practical designing of the EPS main structure is a separate solution of problems on considering the operation reliability of generating equipment and electric networks, whereas it is evident that the required and optimal reliability level in power supply to consumers can be achieved at simultaneous comprehensive solution of both problems. And although the methods for separate solution of these problems comprise corresponding correlations, they are based on approximate relations. Hence, the used methods are effective only for comparatively simple, primarily radial, network schemes [3–7].

Since today, there is no sufficiently accurate and practically convenient integrated optimization model for developing the main structure of complex EPSs. It is suggested that the reliability problems solved now in practical designing should be supplemented by the problem of comprehensive reliability assessment of the formed EPS development variants. The goal of solving this problem is to determine some quantitative indices representing reliability levels of development variants. In fact, the problem of comprehensive assessment is the problem of coordination of all problems separately solved in EPS designing.

2.2 Basic Principles of the Suggested Methodic for Reliability Investigation of Complex EPSs at Their Development Control

2.2.1 Initial Concepts

Estimation of reliability of functioning of complex EPSs, taking into account their physical and technical properties, is proposed to be based on the following concepts.

In general case, the frequency, duration, and level of possible capacity deficits in the system characterize the degree of failure-free operation at power supply. These indicators are determined by amount, composition, and location of generating capacities, network capacity, the failure rate of equipment, power consumption schedule, provision with energy resources (see Fig. 1.1).

In this context, an attempt is made to take into account all or almost all most essential factors in the suggested approach.

The approach is developed based on the following arguments.

1. Within the framework of the stated reliability problem, an electric power system (UPS, IPS, RES) can be treated as a set of consumption (load) nodes, of generating nodes, and of ties between them (transmission lines). Generating nodes and electric ties comprise all basic (generating units and transmission lines) and ancillary equipment of EPS. Each node in such a scheme is concentrated. Ties between the calculation scheme nodes are sets of all transmission lines between the corresponding regions which are substituted for these nodes.

When determining reliability of buses covering the load, the operating schedules of consumers are assumed to be an external factor specified by the power consumption schedule. The EPS states, in this case, can be determined sufficiently fully on the basis of the indicated power consumption schedules, energy parameters, and reliability indices of its basic equipment. Therewith, the reliability indices of basic equipment take into account its outages time (partial and complete) caused by any reasons, such as outages time because of unreliable operation of EPS ancillary equipment (auxiliaries of stations and substations, switchgears, devices of protection, automation, and control). In the context of the accepted concepts, the combinations of states of loads and basic equipment will give insight into EPS basic possible states in space and time.

2. The hierarchical nature of territorial arrangement of EPS allows the analysis to be carried out at each time level stepwise. At first, reliability of the UPS main structure is assessed. Then the obtained data are applied to assess reliability of the schemes calculated at the previous step in less detail (in this case—IPS), and so forth, switching to RES, individual areas within RES, etc.

Therewith, the power supply reliability at each level is understood as reliability of power supply those the buses, which are connection points of equivalent loads in the calculation scheme (feeding substations).

This situation requires coordination of reliability calculations of the EPS main structures at different territorial levels.

The research carried out in this way will ultimately allow us to present a general picture of the reliability of the operation of all links and levels of EPS and, in some cases, will allow us to choose the most rational ways of changing the reliability of the system if the need for such a change becomes apparent.

On the other hand, application of the hierarchical properties of EPS substantially simplifies reliability analysis of a complex power system from the methodological viewpoint and in conclusion allows this analysis to be carried out in sequence and parallel for all the levels with adjustment of the results. The last fact involves time-saving for the similar research.

2.2.2 Calculation Schemes and Calculated States (Modes)

The question of the formation of calculation schemes and calculated states (modes) in the research of the reliability of EPS is important enough, because this determines the correctness of the evaluation of reliability levels. To do this, you can use the methods of equivalention [8–11]. As a result, the schemes of the systems under research decrease the number of calculated nodes and of ties to dimensions not exceeding the dimension of the computational program used, or, if the program does not put this restriction on the basis of the task and purpose of the research.

The level of detail of the settlement schemes is also limited by the time allowed for the research and the possibilities of the applied program used. Other things being equal, the smaller the subsystems in the calculation scheme (in other words, the larger the parts of the EPS included in it), the relatively simpler and faster, though less accurate, analysis of the reliability of this scheme.

Besides, the calculation schemes must be presented in such a way that would provide a required form of assigning initial information for computer calculations and possibility for adjustment of the results obtained and the calculation results at other stages of investigating of the considered EPS and its subsystems.

In making a comprehensive reliability assessment of EPSs, their specific character requires two equivalenting levels of calculation schemes of the investigating system in accordance with two calculated subproblems:

- determination of possible system states and probabilistic characteristics of these states;
- physical and technical analysis of system states determined in the previous subproblem to assess a level and location of possible capacity deficit.

In the first probabilistic subproblem, EPS is considered as a set of the calculated subsystems and ties connecting them. As was mentioned, each subsystem in such calculation scheme is the "concentrated node" with the specified summary load schedule and installed generating capacity. In the probabilistic sense, individual consumers of one subsystem are assumed to be independent and connected in parallel; the generating nodes are represented in a similar way.

The equivalent calculation scheme for the physical and technical analysis of the calculated system states and minimization of possible capacity deficit (an "electric scheme") is constructed based on the initial scheme by equivalenting electric scheme to calculate modes (the aggregation of load nodes, determination of equivalent generators, reduction of the number of calculated ties, etc.). The calculation scheme of the second subproblem must be interrelated and coordinated with the first. In the electric calculated circuit, the ties, which are available in the "probabilistic scheme", must be represented by their substitution (equivalent) schemes.

In the analysis of variants, it is supposed that the generated variants of different EPS structures can be compared. In this context, the calculation schemes of the compared variants for one and the same forecast of EPS development conditions must be identical: contain the same number of the calculated subsystems with identical power consumption conditions; but with different network configurations and parameters, amounts and structures of generating capacities in subsystems.

Considering a multi-level character of EPS as an objective factor and keeping in mind the possibility of relatively arbitrary division of systems into calculated subsystems, sometimes it seems reasonable to generate calculation schemes with account for the formed territorial and organizational hierarchy of EPS. The indicated factor allows practical coordination of the results of reliability calculations at different levels to be provided.

In the EPS, the IPS are accepted as the calculated subsystems in accordance with the above stated as a rule. In IPS, RES, etc., are accepted as the calculated subsystems. Any other ways of partitioning the system into subsystems are possible, depending on the problem being solved. Also, when dividing the system into subsystems, one can take into account the presence of explicit or implicitly expressed "bottlenecks". Occurrence of numerous economic entities governing the energy objects and formation of capacity and electricity markets can demand other principles for construction of calculation schemes.

As noted in Chap. 1, when calculating reliability indicators of electric power supply is understood, that the failure of the system is the transition to any mode (post-accident, planned repair and even in some cases—normal), which is characterized by capacity deficit. It is assumed that the automation and personnel of the system have introduced this regime in the permissible area. Therewith, all available reserves of generating capacity and reserves of communication capacity were rationally used.

As was noted in Chap. 1, when calculating electric power supply reliability indices, a system failure is understood as transition to any condition (postemergency, scheduled repair, and even normal in some cases), which is characterized by deficit of capacity. Therewith, it is assumed that the power system automation and personnel made a deficit state feasible by rationally using all available generating capacity reserves and transfer capability margins of ties.

This section does not deal with disturbances in the work of consumers in emergency transient processes which can be more essential than in the steady postemergency condition. The emergency process duration makes up a small share of the postemergency condition duration, and therefore, some underestimation of the amount of non-delivered electricity to consumers may be considered negligibly small as compared to the undersupply electricity for the whole duration of deficient system state.

This assumption allows one to take into account quite sufficient for practical calculations the operating conditions of the system in the case of non-working states of certain elements. At the same time, analysis is greatly facilitated, since in this case it is not necessary to evaluate the set of all sorts of sudden accidents and prepared outages. And the corresponding transients and conditions that ensure their optimal course, the duration and magnitude of the load shedding depend on many factors. It is very difficult to take into account these factors for complex EPS within an acceptable period even with the help of modern systems analysis tools. In addition, the "Technological Rules for the Operation of Electric Power Systems" declares a requirement for consumers that they themselves should cope with short-term emergency violations of regimes in EPS.

2.3 Conceptual Statement of the Problem, Power System Operation Conditions, and Main Assumptions

One year is usually taken as the calculated time period T_{p} and the problem of determining reliability indices is formulated as follows:

> Determine reliability indices for nodes and the whole system for a year and for intervals of the considered year based on the specified levels and structure of electric power consumption by node (subsystem), network configuration, composition and parameters of EPS equipment (generating units and transmission lines), provision with primary energy resources.

The problem is stated identically for all time stages of designing. The approach to solving this problem is of evaluation character. The calculations aim to determine power supply reliability indices for the whole system and individual areas (energy nodes) for the specified (obtained as a result of performing the corresponding design works) variants of the main structure of EPS of one of the considered territorial levels (UPS, IPS, RES).

The constructed model takes into account the most essential factors of EPS operation which influence its reliability and first of all failures, emergency, and scheduled maintenances of system equipment. Seasonal irregularity of random processes in EPS (e.g., equipment failure flows) and change in a composition and parameters of equipment during the considered period; power consumption schedules in the form of typical daily load schedules including the time zone shifts for different areas of EPS as well as random load deviations and provision with energy resources. Conditions of the calculated system states at the calculated failures are optimized based on the assumed strategy of consumption limitation at individual energy nodes and possible mutual assistance of neighboring energy nodes.

Reliability of complex EPSs in terms of their physical and technical properties is assessed according to the following initial arguments. The considered balance reliability of EPS as a degree of reliable power supply to consumers is characterized by frequency, duration, and level of possible deficit of capacity in the system. When determining power supply reliability with respect to the buses of nodal substations covering load, the working schedules of consumers are taken as an external factor specified by the corresponding equivalent load schedules. The EPS states, in this case, can be determined sufficiently completely based on electric power consumption schedules, energy parameters, and reliability indices of its basic equipment. In the last case, as was already mentioned, consideration should be given to any outages (complete and partial) of the basic equipment including the outages caused by the unreliable operation of ancillary equipment in EPS (auxiliaries of stations and substations, switchgears, protection, automation and control devices), and also the outages due to water shortage at HPPs, fuel shortage at TPPs.

According to the accepted concepts, the general state of loads and basic equipment determines a set of major possible system states in time and space.

The EPS operation model that is based on the indicated initial concepts is characterized by the following features, assumptions, and simplifications.

1. The system failure is understood as an event of EPS transition [12] to any condition which is characterized by deficit of capacity. It is assumed that the automation system and personnel made a deficit state feasible by rationally using all available possibilities for reducing shortages and by limiting consumers by the minimum possible amount.
2. The scheduled maintenances of generating equipment are modeled according to the standards subject to their obligatory fulfillment. The standards can be substituted for the specified schedules of current and major repairs of equipment. It is suggested that the scheduled maintenances of transmission lines should not be considered or considered at the corresponding intervals of the considered year in combination with the emergency repairs (by increasing the relative duration of an outage due to repairs).
3. Equipment failures in the model are not divided into sudden and forecasted. All types of reserves (spinning and standing) are considered to be involved, and the system failure is determined by the total or local load excess over all generating capacity. It is supposed that the available capacity reserve in the system is divided into the spinning and standing reserves according to the ratio between the sudden and forecasted failures. The sudden failures in such a case are covered by the spinning reserve of the first stage (ready reserve) and the forecasted failures with a different degree of lead time—by the spinning reserve of the next stages and by the standing reserve up to their complete utilization.
4. Capacity deficit (correspondingly, electricity undersupply) is determined by the global or local lack of generating capacity.
   The forms of shortage (reduction of frequency in the system, voltages at the buses of consumers, operation of automatic frequency load shedding, etc.) are not dealt with.
5. The features of operating conditions of HPPs, PSPs, and TPPs are considered at the level of special assignment of initial data for these objects (by the corresponding distribution functions of their states in terms of provision with primary energy resources).
6. The maximum or limit transfer capabilities of individual transmission lines are taken constant and independent of system condition, but different (if necessary) for each calculated interval. The total transfer capabilities of ties between the nodes are determined additively as functions of transmission line states (operable and inoperable). However, dependence of transfer capabilities of ties on the state of transmission lines forming them can be set more accurately by the corresponding distribution functions.
7. The electrical modes are not optimized, and capacity deficit in the calculated states are minimized with the help of the interior point (affine scaling) method [13, 14] which is a nonlinear programming method applied to obtain the sought

results with considering the first Kirchhoff law. The second Kirchhoff law approximately takes account of power losses in the internodal ties. The technical and economic characteristics of electricity production and transmission, the characteristics of functioning the wholesale markets of capacity and electricity, the specified strategies of consumer limitation at the specified bilateral limitations on power flows in transmission lines are applied in four suggested optimization models of the calculated states [15]:

- the model, which provides a firm distribution of the total capacity deficit between nodes with considering of losses of capacity in ties (also assessment of only deficient system states);
- the model which makes it possible to minimize capacity deficient in terms of the first Kirchhoff law only and distribute the total system capacity deficit between nodes in proportion to nodal loads (assessment of only deficient system states);
- the model which takes into account technical and economic indicators of expenditures for electricity production and transmission and provides distribution of the total capacity deficit by node in terms of power losses in ties between the nodes (assessment of deficient and not deficient system states);
- the model which is similar to the third one, but in addition takes into account specific features of EPS operation in the context of the wholesale market.

The sets of only postemergency deficient states or all possible states (deficient and not deficient) to take into account fuel consumption and effects of the wholesale markets on EPS operation are analyzed depending on the applied optimization model of deficient capacity. All possible states are considered of course statistically. Their analysis offers additional possibilities for estimation of the distribution functions of transmission lines loading. These distribution functions are the information required for the valid choice of transfer capabilities of ties between the nodes. When optimizing modes of the calculated states, reliability can be studied by considering different means of mutual assistance of electric power companies as a function of the contractual relations between them.

It should be noted that the reliability analysis in the context of EPS survivability and energy security studies calls for calculation of practically all states. This is explained by the fact that in this case consideration is given to heavy disturbances (global fuel undersupply, low-water years, large-scale failures of power plants and major interconnecting ties, etc.).

8. When choosing a reserve of generating capacity, it is taken into account that, on the one hand, each separate energy node should have a certain minimum level of its own reserve, and on the other hand, the overall level of the reservation should be higher than what the calculating technological reserve types (reserve for carrying out of current, major and average repairs, a reserve for carrying out of modernization works, operating reserve).

The minimum level of own reserve is determined by the target reliability standard of power supply to consumers for the electric power company in the case of its isolated functioning. Such standards are known, for example, for the USA ($\mathscr{P}$—probability of the non-deficient operation is no less than 0.9). There is no such standard in Russia. However, the model is adapted for assessment of the needed capacity reserve level in each EPS subject to its isolated operation at the specified $\mathscr{P}$.

This section enumerates the most essential and specific assumptions and simplifications. The reliability theory includes other assumptions and simplifications as well (e.g., independence of equipment failures, neglect of the different failure rate level for operating and repaired equipment, and so on).

The required initial data are:

- the calculation EPS scheme (equivalent energy nodes and ties between them);
- the typical daily load curves at each node, the durations (the number of days of operation) of the corresponding periods, into which the interval (e.g., load of working days and weekends) is divided;
- the mean-square deviations of loads from the forecasted ones;
- the composition of generating units for each node and each considered interval per year;
- the reliability characteristics of basic equipment operation:
  - the duration of major and medium repairs for the indicated EPS elements for a year (standard);
  - the volumes of current repairs for the EPS elements for a year (standard);
  - the probabilities of an emergency outage of generating units;
- the parameters of electric ties between nodes:
  - the probabilities of an emergency outage of transmission lines (1/100 km);
  - transmission capacities of transmission lines in both directions;
  - the length of transmission lines;
- electricity tariffs, full, and variable costs on its production and transmission, if the calculated condition is optimized using the corresponding model, in which the electricity markets are taken into consideration.

In order to optimize the operation of the EPS, taking into account the reliability in interactive mode of the online operator and the computer are required the values of the unit cost in the EPS equipment for all nodes and ties, as well as the compensation costs for all nodes, if under supply of electricity to consumers are required.

It was already noted that when forming the EPS calculation scheme for reliability evaluation, an arbitrary partitioning of the system into subsystems (nodes) is possible. It is recommended to draw up a scheme in such a way that when selecting nodes allocate "weak" ties and take into account the territorial-organizational hierarchy of EPS.

The sought information. As a result of calculations, the following reliability indices are determined for nodes and the whole system for each considered interval (months, quarters) and for the year:

- the probability of failure-free system (node) operation $\mathscr{P}$;
- the average value of electricity undersupply to consumers W_{und};

- the ratios of provision consumer with electricity π;
- the damage D caused by electricity undersupply (in monitory terms with application of the corresponding model of capacity deficit minimization);
- the values of calculated reserves of different types for nodes and the whole system: the reserve for major and medium repairs;
- the reserve for current repairs;
- operating reserve;
- commercial reserve;
- the total volume of reserves of all types.

The model provides for the calculation of the distribution functions of capacity deficits and also the calculation of the energy reliability characteristics (ERChs) of ties.

The energy reliability characteristic (ERCh) of ties here is the distribution function of power flows in the given tie in condition of EPS operation and the specified reliability characteristics of equipment. The indicated function determines a character of interrelation between two adjacent nodes and possible mutual assistance taking from the remaining network.

The considered ERCh is the quantitative reliability characteristic and determines factual effectiveness of using the corresponding tie under the specified conditions. As a result, this characteristic can be applied for mutual coordination of reliability calculations at different territorial levels of EPS.

The dual (objectively conditioned) scarcity estimates of main resources (generating capacity by node and transfer capability by ties) are also determined to support reliability of electricity supply to consumers. These estimates characterize "contribution" of generating capacities of each node and transfer capabilities of each tie to EPS reliability and make possible the reliability optimization in the planned EPS operating modes. In addition, the mean-square deviations of the undersupply of the electric power σW_{und} and coefficient of electricity supply of consumers $\sigma \pi$ are calculated.

2.4 Mathematical Formulation of the Problem on Calculation of Reliability Indices for a Complex EPS

In accordance with the presented problem statement, the EPS calculation scheme is represented as a connected graph, whose vertices (nodes) correspond to equivalent calculated subsystems and edges—to ties between them as a set of transmission lines. The considered period T_{p}, usually equal to year, is divided into S intervals, and at each of them the specified electricity consumption schedules, the composition, and parameters of system equipment are constant. The considered intervals, in turn, can be represented by one or more subintervals (e.g., working days or weekends), each of them being characterized by the constant values of the average loads $P^{1}_{m\varphi}$ at all nodes (taking into account the costs for own needs in the system and losses in distribution networks). Additional information on the calculation scheme construction is presented in Sect. 1.1.

The reliability indices are determined in the following way:

for nodes ($m = \overline{1, M}$):

(a) at the interval $s = \overline{1, S}$

- the probability of failure-free (non-deficit) operation

$$\mathscr{P}_{ms} = 1 - Q_{ms} = 1 - \frac{1}{\tau_s} \sum_{\varphi=1}^{\Phi_s} \sum_{\eta=1}^{H} \sum_{k=1}^{K_s} q_{m\varphi\eta k}^{\text{def}} \cdot \tau_\varphi; \tag{2.1}$$

- the average electricity undersupply to consumers

$$W_{ms}^{\text{und}} = \sum_{\varphi=1}^{\Phi_s} \sum_{\eta=1}^{H} \sum_{k=1}^{K_s} P_{m\varphi\eta k}^{\text{def}} \cdot q_{m\varphi\eta k}^{\text{def}} \cdot \tau_\varphi; \tag{2.2}$$

- the ratio of provision consumer with electricity

$$\pi_{ms} = 1 - W_{ms}^{\text{und}} / W_{ms} = 1 - W_{ms}^{\text{und}} / \sum_{\varphi=1}^{\Phi_s} P_{m\varphi}^{1} \tau_\varphi; \tag{2.3}$$

(b) for the period $T_\text{p} = \sum_{s=1}^{S} \tau_s$ the indices are the same (averaged)

$$\mathscr{P}_{m} = 1 - Q_{m} = 1 - \frac{1}{T_\text{p}} \sum_{s=1}^{S} Q_{ms} \tau_s; \tag{2.4}$$

$$W_{m}^{\text{und}} = \sum_{s=1}^{S} W_{ms}^{\text{und}}; \tag{2.5}$$

$$\pi_{m} = 1 - W_{m}^{\text{und}} / W_{m} = 1 - W_{m}^{\text{und}} / \sum_{s=1}^{S} W_{ms}; \tag{2.6}$$

for the system:

(a) at each interval s the indices $\mathscr{P}_{\text{syst}\,s}$, $W_{\text{syst}\,s}^{\text{und}}$ and $\pi_{\text{syst}\,s}$ are calculated similarly to (2.1)–(2.3);

(b) for the period T_p the indices $\mathscr{P}_{\text{syst}}$, $W_{\text{syst}}^{\text{und}}$ and π_{syst} are calculated similarly to (2.4)–(2.6).

The system indices are calculated based on the determined values of $P_{\text{syst}\,\varphi\eta k}^{\text{def}}$, $Q_{\text{syst}\,s}$ and Q_{syst} which are similar to the values of $q_{m\,\varphi\eta k}^{\text{def}}$, $q_{m\,\varphi\eta k}^{\text{def}}$, $P_{m\,\varphi\eta k}^{\text{def}}$, $Q_{m\,s}$ and Q_m in (2.1)–(2.6).

Here, $Q_{m\,s}$, Q_m, $Q_{\mathrm{syst}\,s}$, Q_{syst} are the relative durations (probabilities) of interruption in power supply to consumers; $\tau_s = \sum_{\varphi=1}^{\Phi_s} \tau_\varphi$ is the interval duration;

φ, Φ_s are the current number of the calculated subinterval at the interval s and the number of the calculated subintervals at the interval s;

τ_φ is the duration of the φ-th subinterval in hours;

η, H are the current number of the calculated random value of irregular load components and the number of their calculated random values;

k, K_s are the current number of the calculated random system state and the number of calculated random system states, which are determined by the k_{ms}-th random values of generating capacities at nodes and the k_{ns}-th random states of transmission lines in ties; $q^{\mathrm{def}}_{m\varphi\eta k}$, $q^{\mathrm{def}}_{\mathrm{syst}\,\varphi\eta k}$ are the probabilities of the capacity deficit $P^{\mathrm{def}}_{m\,\varphi\eta k}$ and $P^{\mathrm{def}}_{\mathrm{syst}\,\varphi\eta k}$ in the $\varphi\eta k$-th calculated state;

$W_{m\,s}$, W_m, $W_{\mathrm{syst}\,s}$, W_{syst} are the required electricity productions.

The capacity deficit probabilities are determined as:

$$q^{\mathrm{def}}_{m\,\varphi\eta k} = \begin{cases} q_{\varphi\eta} \cdot q_k, & \text{if } \Delta P^{\mathrm{opt}}_{m\,\varphi\eta k} > 0, \\ 0, & \text{if } \Delta P^{\mathrm{opt}}_{m\,\varphi\eta k} \le 0; \end{cases} \tag{2.7}$$

$$q^{\mathrm{def}}_{\mathrm{syst}\,\varphi\eta k} = \begin{cases} q_{\varphi\eta} \cdot q_k, & \text{if } \Delta P^{\mathrm{opt}}_{\mathrm{syst}\,\varphi\eta k} > 0, \\ 0, & \text{if } \Delta P^{\mathrm{opt}}_{\mathrm{syst}\,\varphi\eta k} \le 0, \end{cases} \tag{2.8}$$

where $q_{\varphi\eta}$ is the probability of the η-th deviation of random load at nodes at the φ-th subinterval; q_k is the probability of the k-th system equipment state at nodes at the φ-th subinterval; q_k is the probability k-th of equipment state of the system.

The optimal (minimized) volume of the non-served load $\Delta P^{\mathrm{opt}}_{\mathrm{syst}\,\varphi\eta\kappa}$ in the whole system is determined on the basis of the corresponding values of $\Delta P^{\mathrm{opt}}_{m\varphi\eta\kappa}$ in each of M nodes

$$\Delta P^{\mathrm{opt}}_{\mathrm{syst}\,\varphi\eta\kappa} = \sum_{m=1}^{M} \Delta P^{\mathrm{opt}}_{m\,\varphi\eta\kappa}, \tag{2.9}$$

$$P^{\mathrm{def}}_{m\,\varphi\eta k} = \begin{cases} \Delta P^{\mathrm{opt}}_{m\,\varphi\eta k}, & \text{if } \Delta P^{\mathrm{opt}}_{m\,\varphi\eta k} > 0, \\ 0, & \text{if } \Delta P^{\mathrm{opt}}_{m\,\varphi\eta k} \le 0, \end{cases} \tag{2.10}$$

$$P^{\mathrm{def}}_{\mathrm{syst}\,\varphi\eta k} = \sum_{m=1}^{M} P^{\mathrm{def}}_{m\,\varphi\eta k}. \tag{2.11}$$

The values of the calculated random deviations of loads $P^{\mathrm{l}}_{\mathrm{rand}\,\varphi\eta_m}$ from their mean values $P^{\mathrm{l}}_{m\varphi}$ and generation $P^{\mathrm{g}}_{k_{ms}}$ at nodes, and also transfer capabilities of ties $\underline{P}_{k_{ns}}$, $\overline{P}_{k_{ns}}$, $(n = \overline{1, N})$, where n, N are number of the tie and quantity of tie lines between the nodes in the calculation scheme, which are necessary for calculations of $\Delta P^{\mathrm{opt}}_{m\,\varphi\eta k}$, will be determined from the corresponding distribution functions with the help of the Monte Carlo method.

The indicated distribution functions F are assigned or calculated by the specified composition and parameters of the equipment of nodes and ties for each considered interval s:

$$q_{\varphi\eta_m}\left(P^{\mathrm{l}}_{\mathrm{rand}\ \eta_m}\right) = F\left(P^{\mathrm{l}}_{m\,\varphi},\ \sigma^{\mathrm{l}}_m\right),\quad m = \overline{1,\ M} \tag{2.12}$$

$$q_{k_{ms}}\left(P^{\mathrm{g}}_{k_{ms}}\right) = F\left(P^{\mathrm{g}}_{i_{ms}},\ q_{i_{ms}}\ /\ i_{ms} = \overline{1, I_{ms}}\right),\quad m = \overline{1,\ M}; \tag{2.13}$$

$$q_{k_{ns}}\left(\underline{P}_{k_{ns}},\ \overline{P}_{k_{ns}}\right) = F\left(\underline{P}_{i_{ns}},\ \overline{P}_{i_{ns}},\ q_{i_{ns}}\ /\ i_{ns} = \overline{1, I_{ns}}\right),\quad n = \overline{1,\ N}, \tag{2.14}$$

where σ^{l}_m is mean-square load deviation in p.u. from the values of $P^{\mathrm{l}}_{m\varphi}$ (the distribution law is assumed to be normal);

$P^{\mathrm{g}}_{k_{ms}} = P^{\mathrm{g}}_{\mathrm{avail}\,ms} - P^{\mathrm{g}}_{\mathrm{emerg}\,k_{ms}}$ is the calculated k_{ms}-th value of the total generating capacity which is not in emergency outage;

$P^{\mathrm{g}}_{i_{ms}}$, $q_{i_{ms}}$ are available capacity and probability of the emergency outage of the i_{ms}-th units;

i_{ms}, I_{ms} are current number of the unit and quantity of unit (or the calculated steps of generating capacity);

$\underline{P}_{i_{ns}}$, $\overline{P}_{i_{ns}}$ are transfer capability limits of the i_{ns}-th transmission line in the directions taken as negative and positive (inverse and direct), respectively;

$q_{i_{ns}}$ is outage probability of the i_{ns}-th transmission line:

$$q_{i_{ns}} = \begin{cases} q_{\mathrm{emerg}\,i_{ns}} & \text{without regard to scheduled repair transmission lines;} \\ q_{\mathrm{emerg}\,i_{ns}} + q_{\mathrm{sched}\,i_{ns}} & \text{with regard to scheduled repair transmission lines,} \end{cases}$$

where $q_{\mathrm{emerg}\,i_{ns}}$, $q_{\mathrm{sched}\,i_{ns}}$ are outage probabilities of the i_{ns}-th transmission line in the emergency and scheduled maintenances, respectively; i_{ns}, I_{ns} are current number of the transmission line and total quantity of transmission lines in the tie.

The optimal values of capacity deficit $P^{\mathrm{def}}_{m\,\varphi\eta k}(\Delta P^{\mathrm{opt}}_{m\,\varphi\eta k})$ are determined by optimizing the conditions of calculated EPS states as a function of the accepted strategy on consumer limitation, conditions of the electricity market. In the simplest case [15–18] for capacity deficit distribution by node proportionally to the load power, the following function is minimized:

$$\sum_{m=1}^{M} C_m \Delta P^{\mathrm{l}}_{m\,\varphi\eta k}, \tag{2.15}$$

where

$$\Delta P^{\mathrm{l}}_{m\,\varphi\eta k} = P^{\mathrm{l}}_{m\,\varphi} + \Delta P^{\mathrm{l}}_{\mathrm{rand}\,\varphi\eta_m} - P^{\mathrm{l}}_{m\,\varphi\eta k} > 0 \tag{2.16}$$

if the following conditions and constraints are met to obtain the solutions admissible in terms of the physical and technical grounds

$$P^{\mathrm{l}}_{m\,\varphi\eta k} - P^{\mathrm{g}}_{m\,\varphi\eta k} + \sum_{n=1}^{N_m} P_{n\,\varphi\eta k} = 0 \tag{2.17}$$

for $m = \overline{1, M}$; $\varphi = \overline{1, \Phi_s}$; $\eta = \overline{1, H}$; $k = \overline{1, K_s}$; $s = \overline{1, S}$.

Here C_m are coefficients determining importance of load in node; $P^{\mathrm{l}}_{m\,\varphi\eta k}$, $\Delta P^{\mathrm{l}}_{m\,\varphi\eta k}$ are values of the covered and uncovered load, respectively; $P^{\mathrm{g}}_{m\,\varphi\eta k}$ is value of the utilized generating capacity; $P_{n\,\varphi\eta k}$ is capacity flow in the tie (the power flow from the given node to the neighboring ones is taken here as a positive direction and its value is used with the sign plus, the power flow from the neighboring nodes to the given one—as a negative direction and its value is used with the sign minus); N_m is number of the tie lines adjoin to the given node.

Components of the balance equations (2.17) are determined as follows:

$$0 \le P^{\mathrm{l}}_{m\,\varphi\eta k} \le P^{\mathrm{l}}_{m\,\varphi} + P^{\mathrm{l}}_{\mathrm{rand}\,\varphi\eta_m}; \tag{2.18}$$

$$0 \le P^{\mathrm{g}}_{m\,\varphi\eta k} \le P^{\mathrm{g}}_{\mathrm{work}\,k_{ms}},$$

where

$$P^{\mathrm{g}}_{\mathrm{work}\,k_{ms}} = P^{\mathrm{g}}_{k_{ms}} - P^{\mathrm{g}}_{\mathrm{sched}\,ms},\ k_{ms} = \overline{1, K_{ms}}; \tag{2.19}$$

$$\underline{P}_{k_{ns}} \le P_{n\varphi\eta k} \le \overline{P}_{k_{ns}},\ \text{where}\ k_{ns} = \overline{1, K_{ns}}. \tag{2.20}$$

Here $P^{\mathrm{l}}_{\mathrm{rand}\,\varphi\eta_m}$ is taken from expression (2.12), $P^{\mathrm{g}}_{\mathrm{work}\,k_{ms}}$ is working capacity; $P^{\mathrm{g}}_{\mathrm{sched}\,ms}$ is capacity in the scheduled repair.

The available capacity $P^{\mathrm{g}}_{\mathrm{avail}\,ms}$ at each node (for calculation of $P^{\mathrm{g}}_{k_{ms}}$) at each interval s is determined based on the specified initial information about equipment:

$$P^{\mathrm{g}}_{\mathrm{avail}\,ms} = \sum_{i_{ms}=1}^{I_{ms}} P^{\mathrm{g}}_{i_{ms}}. \tag{2.21}$$

The scheme of accounting the scheduled repairs of generating facilities $P^{\mathrm{g}}_{\mathrm{avail}\,ms}$ is determined in the following way:

$$P^{\mathrm{g}}_{\mathrm{sched}\,ms} = P^{\mathrm{g}}_{\mathrm{curr}\,ms} + P^{\mathrm{g}}_{\mathrm{major}\,ms}; \tag{2.22}$$

$$P^{\mathrm{g}}_{\mathrm{curr}\,ms} = \sum_{i_{ms}=1}^{I_{ms}} \bar{\alpha}_{\mathrm{curr}\,i_{ms}} \cdot P^{\mathrm{g}}_{i_{ms}}; \tag{2.23}$$

$$P^{\mathrm{g}}_{\mathrm{major}\,ms} = f_1(V_{\mathrm{major}\,m}, F_{\mathrm{valley}\,m}); \tag{2.24}$$

$$V_{\text{major } m} = \left(\sum_{i_m=1}^{I_m} \bar{\tau}_{\text{major } i_m} \cdot P^{\text{g}}_{i_m} \right) / k_{\text{valley}}\,; \tag{2.25}$$

$$F_{\text{valley}\,m} = f_2\left(P^{1}_{\max\, ms}, s = \overline{1, S}\right). \tag{2.26}$$

The current repairs $P^{\text{g}}_{\text{curr}\,ms}$ (2.23) and correspondingly the major and medium repairs $P^{\text{g}}_{\text{major}\,ms}$ (2.24)–(2.26) are taken into account independently using different approaches [4]. The common feature of accounting both types of repairs in the model is the absolute provision of the required volumes of these repairs.

In formulas (2.23)–(2.26):

$\overline{\alpha}_{\text{curr}\,i_{ms}}$ is the standard (relative total duration) of the scheduled current repairs of the i_{ms}-th unit; $V_{\text{major}\,m}$ is the required "area" of the major repairs of generating equipment at the node (MW·day); $F_{\text{valley}\,m}$ is the area of the valley in the monthly peak load schedule of the node (MW·day); $\bar{t}_{\text{major}\,i_m}$ is the standard of the scheduled major and medium repairs of the i_m-th unit (days/year); k_{valley} is the capacity factor of the valley in the annual schedule of monthly peak loads for major repairs; $P^{1}_{\max\,ms}$ is the peak load at the node.

Additional information obtained as a result of solving the formulated problem comprises probabilities of capacity deficit of different levels [formulas (2.7), (2.8)]; probabilities of the power flows $P_{n\varphi\eta k}$ (energy reliability characteristics—ERCh) of the ties, which are obtained by solving problem (2.15)–(2.20); dual estimates obtained by optimization of capacity deficit in power systems.

This information allows us to calculate the root-mean-square deviation (rms deviation) of the exponents W_{und}, π.

The series of distribution of capacity deficits by nodes and in the system as a whole for each s-th interval and for the entire calculation period T_{p}: $q^{\text{def}}_{ms}(P^{\text{def}}_{ms})$, $q^{\text{def}}_{m}(P^{\text{def}}_{m})$, $q^{\text{def}}_{\text{syst}\,s}(P^{\text{def}}_{\text{syst}\,s})$, $q^{\text{def}}_{\text{syst}}(P^{\text{def}}_{\text{syst}})$, with the known frequency coefficient, which determines the dependence of the load on the frequency, transformed depending on the operation times of EPS with different frequencies below 50 Hz: $\tau_f = F(f)$ for the system as a whole and its individual nodes in each s-th interval and for T_{p}. This information also makes it possible to determine the ERCh of the ties and the integral dual estimates for each node and each the tie.

The problem of reliability assessment of a large scheme is a complex "tree" of subproblems. As a result of their solution the values of $q^{\text{def}}_{m\,\varphi\eta\kappa}$ and $P^{\text{def}}_{m\,\varphi\eta\kappa}$ are determined as the basis for calculation of reliability indices.

2.5 Methods and Techniques for Realization of the Stated Problem Solution

The calculation scheme of EPS reliability indices is represented as a sequence of the following blocks:

I. Block of initial data preparation:

1. Initial data input, analysis, and processing.
2. Calculation of complex characteristics of the factors determining reliability of nodes and the whole system.

II. Probability block.

3. Construction of load schedules and schedules of planned repairs of generating equipment for each node.
4. Calculation of distribution functions of states of generating equipment for each node.
5. Determination of possible states of electric ties taking into account of transmission line idle times in repairs.
6. Generation of calculation states (modes) of generating equipment, transmission lines and load (considering its random fluctuations) in the system.

III. Block of optimization of calculated state conditions (modes):

7. Choice of the optimization model of calculated state conditions.
8. Optimization of the calculated state condition, including minimization of capacity deficit (determination of capacity deficit by node, economic indices and capacity flows in ties).

IV. Block of calculation of reliability indices:

9. Calculation of energy reliability characteristics for the time intervals s and the considered period T_{p}.
10. Calculation of nodal reliability indices for the time intervals and the considered period T_{p}.
11. Calculation of system reliability indices for the time intervals and the considered period T_{p}.
12. Calculation of economic indices of the system for the time intervals and the considered period T_{p} (as a function of the model chosen in block III).
13. Processing and presentation of calculation results.

Some subproblems are included in the general problem as reusable algorithms. For example, stages 3–12 are calculated for all intervals S.

A different character of the subproblems solved when calculating the reliability indices stipulates application of a set of different methods, which comprise:

- the method for calculation of distribution series of random states of the calculated system scheme elements on the basis of the generating function of the general theorem on experiment recurrence and the theorem on addition and multiplication of probabilities of different events;
- the Monte Carlo method for generation of the calculation system states;
- the combinatorial methods for separation of system states by the required attribute;
- the linear and nonlinear programming methods for the optimization problem of calculated system states (the interior point method [13]).

Some of the indicated methods are known and have been applied to the problems for long, the others are somewhat modified or are the authors' original developments. The next chapter and Appendix A are devoted to detailed presentation of the methods for problem solution and the corresponding algorithms.

In order to realize algorithms of the suggested methods, a high-dimensional problem of sets of system states can be solved by considering a hierarchical structure of EPSs.

The problem is solved by the simulation modeling method of EPS operation for the considered period. The states of system loads and equipment are determined by the Monte Carlo method. The distribution series of unit capacities at nodes and transfer capabilities of transmission lines in ties, which are in emergency outages, are calculated preliminary.

Advantages of the suggested and implemented algorithm over the known ones, which are briefly described in Malkin and Kovalev [19], consist in the following.

- Possibility for research systems of complex configuration with the ties of limited transfer capability.
- Preliminary equivalenting of equipment composition is not required for calculations. This fact reduces laboriousness of initial data preparation and improves calculation accuracy, particularly for EPSs with a large number of diverse units.
- To make the program universal, in the reliability calculations it is foreseen to apply both the specified distribution functions of equipment states by node and tie and the functions calculated in the program using the data on reliability of individual system elements.
- The program allows the initial information to be set in different forms and with a different degree of accuracy.
- The complex indices which characterize properties and conditions of EPS operation in a generalized way and determine its reliability are calculated in addition to the reliability indices and the parameters of their scattering (see Fig. 1.1).
- Optimization of the calculated states (modes) in terms of operational strategy of consumer limitation allows the reliability indices of electric power supply having a real sense to be obtained for the whole system and its individual nodes.
- Possibility for assessment of reliability indices for the whole considered period and its individual intervals.
- In parallel with the dual estimates by node and tie, the energy reliability characteristics of ties are used as information for the technical and economic analysis of efficient operation of the research system.
- The program has a block structure which makes it possible to change its rather easily in the improvement or modification process.

In the general case, the number of possible calculated nodes and ties depends on capabilities of computational tools (their speed and the volume of random-access memory, but one should take into consideration that the up-to-date tools have in this context *very* high indices offering *very* great opportunities), as well as on the efficiency of program algorithms. In particular, this is manifested in the choice of the method of the repeatedly working block of minimizing the capacity deficit.

Almost until 1990, there were very large limitations on speed and size of RAM, the number of possible calculated nodes and ties in the calculation scheme was determined by the admissible time of one calculation on the computational tool (no more than 5–6 h for the considered problem). On the other hand, the calculation time depended on the computer speed. However, as far as the main calculation time depends on the analysis of the calculated system states, it was suggested that the sizes of the limiting calculation scheme should be determined by some number of states G which can be analyzed by the program for the desired calculation time.

On the basis of the expression

$$G = \sum_{s=1}^{S} \left(\prod_{m=1}^{M} K_{ms} \cdot \prod_{n=1}^{N} K_{ns} \cdot \prod_{m=1}^{M} B_m \cdot \Phi_s \right) \cdot K_{\mathrm{scr}}, \tag{2.27}$$

and the known average calculation time t_{calc} of one system state, which in the general case is the function of the number of nodes and ties in the scheme, it is possible to estimate:

- the approximate time of the whole calculation at the specified $S,\ M,\ N,\ K_{ms},\ K_{ns},\ B_m$ and Φ_s

$$T_{\mathrm{calc}} = t_{\mathrm{calc}} \cdot G; \tag{2.28}$$

- the limiting values of $S,\ M,\ N,\ K_{ms},\ K_{ns},\ B_m$ at the predetermined limitation on the calculation time, as far as

$$G = T_{\mathrm{calc}}/t_{\mathrm{calc}}. \tag{2.29}$$

Here s, S are current number of the considered interval and quantity of these intervals; n, N are current number of the tie and quantity of ties in the calculation scheme of EPS; m, M are current number of the node and quantity of nodes in the calculation scheme; K_{ms} is quantity of the calculated random states of generating capacity at the m-th node of the s-th interval for different composition of equipment; K_{ns} is quantity of the calculated random states of transmission lines (the calculated values of transfer capabilities) of the n-th tie at the s-th interval for different transmission line compositions; B_m is quantity of the calculated random values of irregular load components at the m-th node; Φ_s is quantity of the considered periods at the s-th interval which are determined by the constant values of the regular load $P^1_{m\varphi_s}$ at all nodes; K_{scr} is screening coefficient of shortage-free states which is determined by the expression $K_{\mathrm{scr}} \approx 1 - \mathscr{P}_{\mathrm{syst}}$ (for details see Appendix A.1.9).

As was already mentioned, the issue of designing the calculation scheme and calculated conditions for the reliability analysis of particular EPSs is an expert issue and depends on possibilities of different assignment of parameters for the calculated conditions and dimension of the system scheme. These parameters can be varied by changing the indicated above variables $S,\ M,\ N,\ K_{ms},\ K_{ns},\ B_m$ and Φ_s etc.

The use of energy reliability characteristics (ERCh) of electric ties (see Sect. 3.2.11) allows the reliability for the scheme of any size to be analyzed by a series of sequential calculations with a sufficient accuracy. For this purpose, the scheme representing the whole system in an aggregated form, practically from 2 to 5–8 nodes (8—number of IPSs in UPS of Russia), is calculated at the first stage. The energy reliability characteristics (ERCh) of ties between the territorial regions (zones) of EPSs assumed as calculated nodes are determined in addition. At the second stage, each previous node is represented as an independent calculation scheme with the corresponding number of subsystems and ties between them. The joint operation nodes with the system are taken into consideration based on the energy reliability characteristics (ERCh) calculated at the first stage. The calculation scheme sizes at the second stage as well as at the first stage are determined by using formulas (2.27)–(2.29). The generating capacity states of subsystems are determined subject to the energy reliability characteristics of ties in these subsystems with the rest of the system.

The described sequence of calculation of complex schemes is performed until the reliability characteristics of the required territorial levels are determined.

The sufficient calculation accuracy of reliability indices which is necessary for the constructed model to provide a specified comprehensive reliability assessment of the bulk power system development scenarios is achieved by:

- taking into account in the model all the most significant factors affecting reliability;
- representing these factors practically in their most complete and exact form;
- the accurate work of the main computational modules of the software (for details see Chap. 3 and Appendix A);
- the required format and accuracy of initial information representation in the considered computational model (see Chap. 4).

References

1. Avramov NN, Bartosh L, Grabak J et al (1988) Reserves of capacity in the electric power systems of the CMEA member countries. Methods of research. Science, Siberian Branch, Novosibirsk, 150 p (in Russia)
2. Dikin II (1978) The method of interior points in mathematical programming. In: Applied mathematics. Science, Novosibirsk, pp 133–158 (in Russia)
3. Methodical instructions for designing the development of power systems (2003). Izd-vo NTs ENAS, Moscow (in Russia)
4. Rokotyan SS, Shapiro IM (eds) (1985) Handbook on the design of electric power systems. Energoatomizdat, Moscow, 352 p (in Russia)
5. Guk YB, Losev EA, Myasnikov AV (1974) Evaluation of the reliability of electrical installations. In: Konstantinova BA (ed). Energia, Moscow, 200 p (in Russia)
6. Sinchugov FI (1971) Calculation of the reliability of electrical connections schemes. Energia, Moscow, 176 p (in Russia)
7. Volkov GA (1968) Determination of the optimal reserve of active power in the unification of power systems, izvestiya AN SSSR. Power Transp 4:79–82 (in Russia)

8. Rudenko YN, Cheltsov MB (1974) Reliability and redundancy in electro-energy systems. Methods of research. Science, Siberian Branch, Novosibirsk, 264 p (in Russia)
9. Rozanov MN (1984) Reliability of electric power systems. Energoatomizdat, Moscow, 200 p (in Russia)
10. Kozlov BA, Ushakov IA (1975) Handbook on calculating the reliability of radioelectronics and automation equipment. Soviet Radio, Moscow, 472 p (in Russia)
11. Voropai NI, Gamm AZ, Krumm LA et al (1970) On the problem of equivalence in the construction of mathematical models. In: Optimization and control in large energy systems. Tom. 1. SEI SB AS USSR, Irkutsk, pp 193–218 (in Russia)
12. Reliability of energy systems (a collection of recommended terms) (2007). IAC "Energia", Moscow, 192 p (in Russia)
13. Dikin II, Zorkal'tsev VI (1980) Iterative solution of problems for mathematical programming (algorithms for the method of interior points). Science, Novosibirsk, 148 p (in Russia)
14. Kovalev GF, Lebedeva LM (2000) A complex of models for optimizing the modes of calculated states in assessing the reliability of electric power systems. Preprint. ISEM SB RAS, Irkutsk, 73 p (in Russia)
15. Troshina, GM (1978) On an approach to the solution of the problem of minimizing the deficiency of power in electric power systems. In: Methodological issues of investigating the reliability of large power systems, issue 15. SEI SB AS USSR, Irkutsk, pp 34–43 (in Russia)
16. Zorkaltsev VI, Kovalev GF, Lebedeva LM (2000) Models for estimating the power deficit of electric power systems. Preprint. ISEM SB RAS, Irkutsk, 25 p (in Russia)
17. Dikin II (1969) Convergence of an iterative process. In: All–union conference on problems of theoretical cybernetics (theses of reports). Science, Novosibirsk, pp 56–57 (in Russia)
18. Dikin II (1978) Application of algorithms for the method of internal points to minimize the power deficit in the electric power system. In: Methodological issues of research on reliability of large energy systems. Academy of Sciences of the USSR, SEI, Irkutsk, pp 23–33 (in Russia)
19. Malkin PA, Kovalev GF (1981) Software for the calculation of reliability in the design of EPSs. In: Methods for Investigating the reliability of large energy systems, issue 22. SEI SB AS USSR, Irkutsk, pp 7–16 (in Russia)

Chapter 3
Programs for Analyzing the Reliability of EPS, Conditions, and Basic Provisions of Their Application to the Design Practice

3.1 Key Requirements for the Programs and Conditions of Their Realization

This chapter presents two programs. One of them is "KORALL," which is the originally created program by the suggested methods (1972–1976). The second programm is "YANTAR" is the evolution and improvement of the "KORALL" program both methodically and in terms of the progress of computer technology.

The "KORALL" program was applied for all initial calculations and research, which show adequacy of the suggested methods at the moment of program design, but correspond to the then development level of computer technology (limited capabilities of the random-access memory (RAM), high speed, etc.) and also for comparison of the results obtained on the basis of other similar programs. The difference between the "KORALL" and "YANTAR" programs will be not described in the work. Note that all main methods developments were included in the "YANTAR" program to a lesser or greater degree. The "KORALL" program is mentioned only because of importance of the research performed on its base, which gives answers to the question on the robustness of the suggested approach, its adequacy despite such a long time span.

The problem of reliability assessment of a large power system such as UPS, IPS, or RES represents, as was already said, a "tree" of subproblems, part of which is included in the general problem as reusable, but complicated algorithms.

The requirements of program text compactness, calculation efficiency, minimization of the amount of RAM used, simplicity of making changes in the program text by the developer, automation of variant-to-variant transition during mass calculations, detection of errors and malfunctions, self-control of the calculation process, possibility for calculation interruption and recommencement, and some others were central at the moment of the program design. It should be noted that in principle, most of the requirements remain actual despite the rapid progress in computer technology.

The use or availability of data banks of the input and output information is one of the most important requirements earlier and now.

G. F. Kovalev and L. M. Lebedeva, *Reliability of Power Systems*,
Power Systems, https://doi.org/10.1007/978-3-030-18736-1_3

It was decided to create a complex, adhering to a block (module) structure, and in the FORTRAN programming language. It in contrast to other programming languages, it is best suited for solving technical problems in the context of work with RAM and performance of computational operations with a high accuracy, particularly in simultaneous work with the information changing in the range from $\pm 10^{-12}$ to $\pm 10^{12}$ and more.

The model was designed to assess the reliability of the basic structures of EPS for different territorial levels and the time stages of their development.

3.2 Specific Features of the Algorithms and Programs for Calculation of Reliability Indices of a Complex EPS (The "KORALL" and "YANTAR" Programs)

The designed programs are an effort to apply the best characteristics of available algorithms tested in practice. In the presented programs, the reliability indices are calculated on the basis all those factors (in their most complete and exact form), that are treated as main, but not considered simultaneously or sufficiently accurately in other programs.

At the same time, the algorithm of solving individual subproblems includes new modifications developed by the authors, which are intended to decrease computing time or improve the accuracy of the corresponding calculations.

This section is devoted to describing only the features of the software implementation of the problem of calculating reliability indicators of complex EPS (on the example of the program "YANTAR") in comparison with known implementations (see Chap. 1).

Figure 3.1 presents the block diagram of the "YANTAR" program. Peculiarities of individual sub-blocks are presented in the sequence of their placing in the block diagram. The algorithms of program work are fully described in Appendix A.

3.2.1 Initial Data

Figure 3.2 demonstrates a basic form of the calculation scheme. In this context, the key feature of the "YANTAR" program is the possibility for research systems of complex configuration with the ties of limited transfer capability and possible failures of transmission lines forming them.

The internal structure of nodes is a detailed representation of energy areas, which are taken as "concentrated" at a higher hierarchical level (FFA—flow-free area).

A set of initial information was somewhat expanded in the "YANTAR" program due to consideration of losses in ties and market relations in the electric power industry.

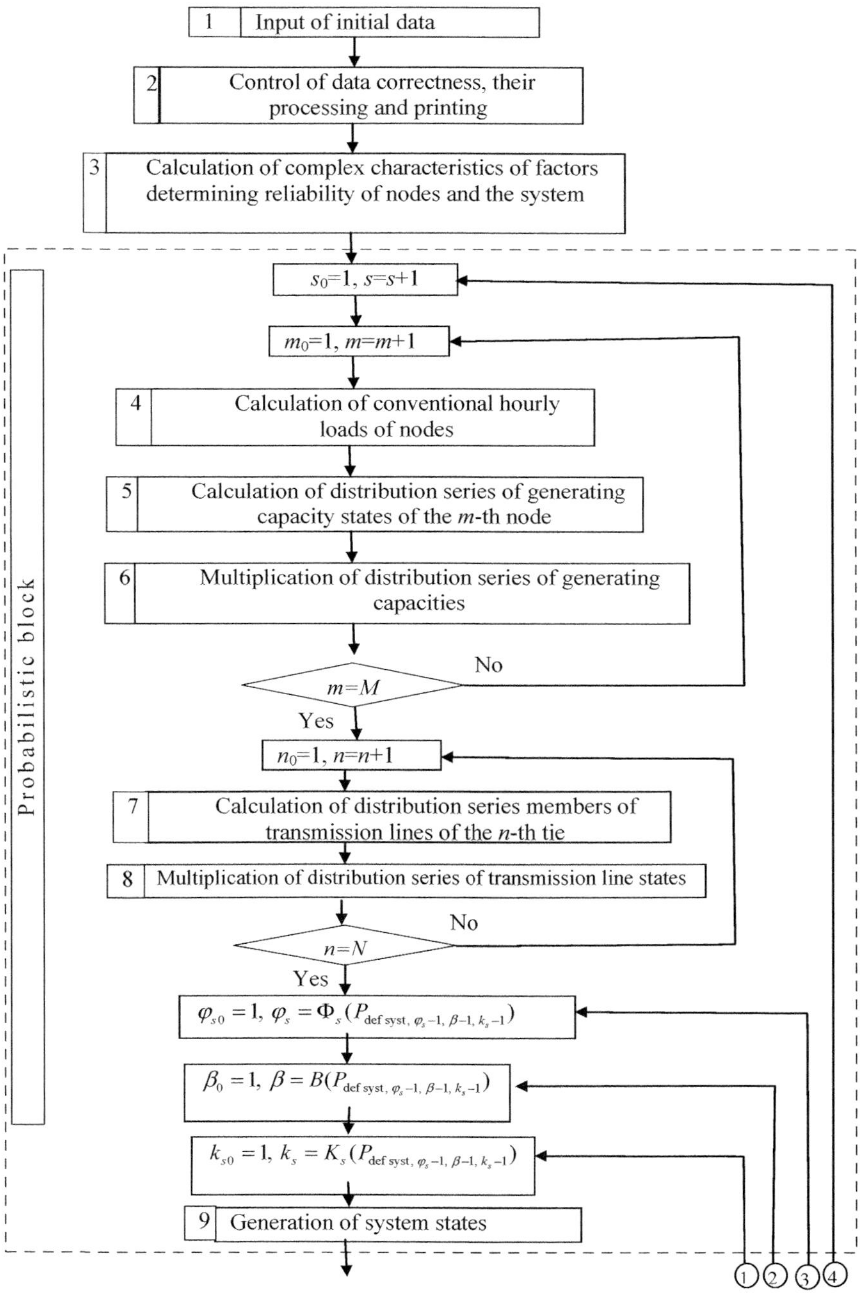

Fig. 3.1 Block scheme of the "YANTAR" program for determination of reliability indices of a complex EPS

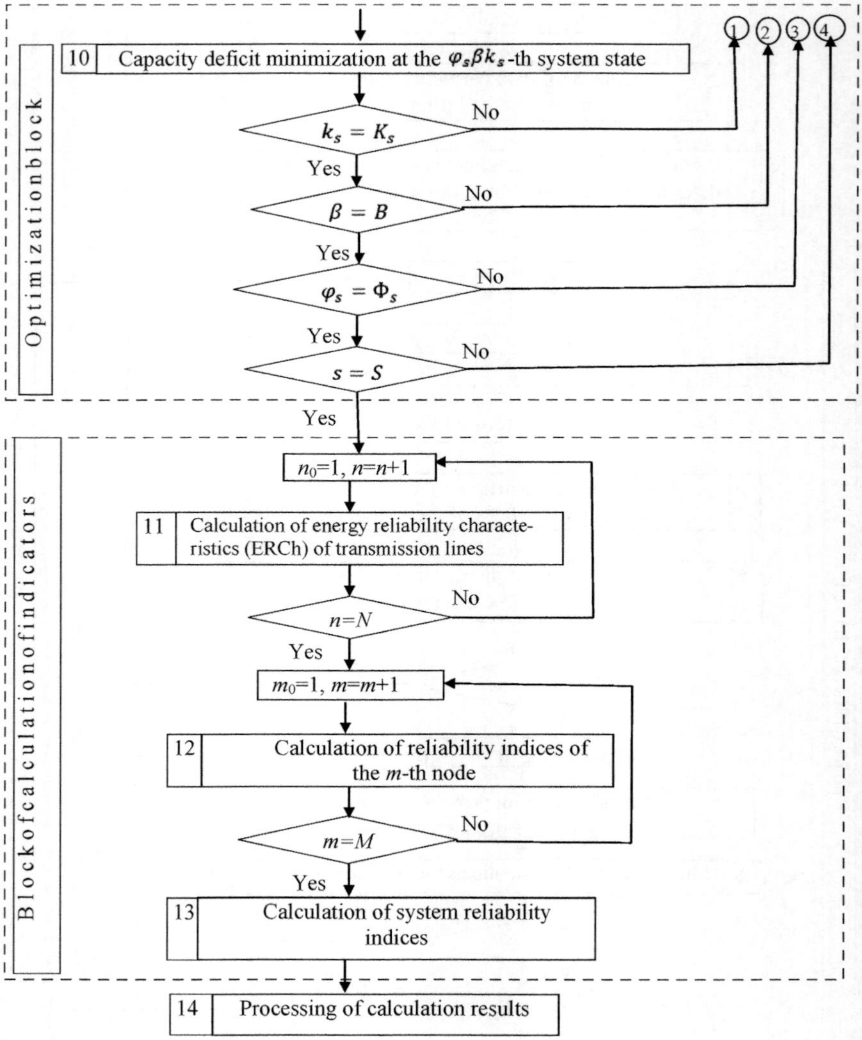

Fig. 3.1 (continued)

When designing individual subprograms, the authors paid attention to simplicity and flexibility of forms of assigning the most common initial data on loads and equipment for design practice. The subroutines of the program "YANTAR" provide the ability to specify the initial data in various forms and with varying accuracy, which may be required for research calculations and calculations in design practice (see Sect. 1.6), considering:

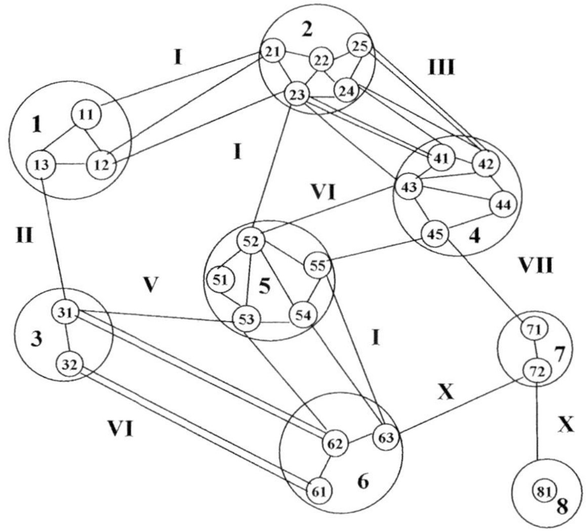

Fig. 3.2 An example of the electric power system (EPS) calculation scheme (the amount of nodes $M = 8$, the amount of ties $N = 11$)

- different requirements to the accuracy of calculations in solving the problems of managing the development of EPS;
- the need to research the influence of factors that determine reliability, on the sought-for indicators;
- investigating the impact of the accuracy of the representation of the source information on the calculated indicators.

For generating equipment, it is possible to change its composition at the considered period (the information is entered for each interval s). This makes it possible to consider changes in the parameters of equipment composition and the composition itself, including equipment commissioning and decommissioning.

The installed equipment set is specified as groups of units. Here, the group is understood as a body of homogeneous units of the same capacity, the same repair standards, and the same emergency rates. Specifying of the information on units as groups is sufficiently flexible and, thus, allows a rather easy analysis of many factors influencing the power system operation reliability, for example, equipment running-in, differences in the repair standards for units of the same capacity, and also provision with energy resources (in the form of its assigned probability b_{ims}).

To make the algorithm universal, the parts of generating capacity of the node or the total generating capacity can be set by the distribution series of states. In the case of necessity, it is possible to use not the distribution series obtained with the help of the equipment unavailability factor, q_i^g, but the factor obtained by some other (probably more accurate) means or, suppose, by the statistical modeling method to estimate an equipment state. Besides, it also becomes possible to assign reliability characteristics of the two-boiler single-turbine units more accurately and to take into account energy reliability characteristics (ERChs) of external ties adjacent to the node, etc.

The nodal loads are set as typical daily schedules (in p.u.), the difference in time zones being taken into consideration. The values of regular annual peak loads (in MW), the monthly peak loads in p.u. from the annual peak load, and the daily peak loads in p.u. from the monthly peak loads can also be set by node.

The random ("irregular") load fluctuations are simulated by setting the distribution function (the normal law represented by the stepwise approximation (in p.u.) with the specified mean-square deviation (msd) σ^1).

Load schedules should include loads of consumers, own needs of stations, and substations, as well as capacity losses in the elements of energy nodes.

For transmission lines, just as for generating equipment, the data on the set and parameters of transmission lines are defined for each calculated interval to take into account their change during the year. Change in the emergency rate and the frequency of scheduled repairs by season of the year is particularly topical for transmission lines.

When setting the information on transmission lines, their scheduled repairs can be taken into consideration jointly with the emergency repairs and can be assigned by the average total relative specific duration of their outages in the scheduled $q_{0\,i_{ns}}^{\text{sched}}$ and emergency $q_{0\,i_{ns}}^{\text{emerg}}$ repairs

$$q_{0i_{ns}} = q_{0i_{ns}}^{\text{sched}} + q_{0i_{ns}}^{\text{emerg}} \quad 1/100\ \text{km}.$$

If the system scheme contains transmission lines of complex configuration (with switching stations and sections with different parameters, etc.), the information on such transmission lines is defined either in the form of equivalent transmission lines of the length $H_{i_n}^{\text{equi}}$, by the specific duration of outage $q_{0i_{ns}}$, etc., or in the form of the distribution series of states of these complex transmission lines.

To make the algorithm universal, if required, the distribution series of states of the groups of transmission lines can be assigned instead of their parameters $q_{0i_{ns}}$, as is the case for generating capacity.

3.2.2 *Setting, Accuracy Control, Generation, and Processing of Initial Data Array*

In addition to error detection within the frame of the Fortran language (correspondence of the identified arrays to the assigned ones, availability of the required initial data), the program is able to detect errors with the help of individual logical tests. If the errors are detected, the information on them is written in the special file and displayed (for further details, see Appendix A.2).

3.2.3 Calculation of Complex Characteristics of the Factors Determining Reliability of Nodes and the Whole System

The given information is used to determine some variables required for calculation of reliability indices and indices, which characterize the factors influencing reliability in a generalized way. The generalized indices are needed for research purposes and practical analysis of the impact of operation conditions of the considered system on its reliability. The corresponding indices are presented in Appendix A.3.

3.2.4 Calculation of Conditional Hourly Loads of Power Nodes

The problem is to calculate schedules of conditional hourly loads of power nodes load, which include power consumption and also capacities of generating equipment in the scheduled repairs. The problem is solved similarly for each node in the following way:

(1) Schedules of major repairs for each month are calculated according to the annual schedule of monthly highs, the composition of the units at the beginning of the year, and the corresponding standards for repairs.
(2) Schedules of current repairs for intervals s are calculated on the corresponding equipment compositions and repair rates.
(3) For the given daily load schedules of each s-th interval, the capacities of current and major repairs are added to each hour.

The corresponding computational algorithms are based on the following initial assumptions:

(a) The schedule of major repairs for a year is made up only for the equipment specified by the beginning of the year.
(b) The major repairs of basic equipment are planned on the assumption that they should be performed primarily during seasonal load curtailment in the system.
(c) The capacities for current repairs are determined on the assumption that they are performed *uniformly* during the year based on the standards (percent of the installed capacity of different types of units).

3.2.5 Calculation of the Distribution Series of Generating Capacity States for a Separate Node

Calculation of the distribution function (series) of generating capacity states is a necessary operation in assessing power system reliability. These calculations are complex and time-consuming, and the complexity increases with the development of an EPS, as the number and variety of units installed in the system increases, the number of which sometimes reaches several hundreds.

On the other hand, the programs, which include algorithms for calculation of distribution functions of generating capacity, impose general requirements for the use of RAM, maximum speed, and reasonable accuracy of calculations.

Thus, the generation of the corresponding effective algorithm was and remains a topical problem, which always needs much attention when designing the reliability assessment programs.

Such complex and time-consuming calculations of the sought function members necessitated the elaboration of different techniques, which facilitate and simplify the considered procedure. Equivalention is recommended in publications as a main simplifying technique. Several different ways of presenting the real generating elements of the system by a group of similar equivalent units have been proposed [1, 2]. Transition to the calculated unit with the equivalent capacity and emergency rate made it possible to effectively apply the binomial distribution formula and even the Poisson law to calculate members of the series [2].

As noted in Sect. 1.7, some authors of the programs used an even deeper equivalence, replacing the generating function of the theorem on the repetition of experiments by a normal law [3, 4] or the Charlier and Pierson law [5–7].

However, the equivalenting of any level gives the sought result with an error [1, 8], which sometimes is quite significant (for further details, see Sect. 4.3.3).

In parallel with searching for the correct methods of equivalenting, it seems feasible to solve the problem without preliminary representation of real generators as equivalent ones, taking advantage of the capabilities of up-to-date computers.

For this purpose, it is possible to apply the Monte Carlo method or the so-called "analytical" method that uses the formulas known from the classical probability theory. The algorithm for calculation of the distribution function of generating capacity states of the power system with practically any amount of units of different parameters is described in Appendix A.

The designed algorithm is characterized by some specific features: feasibility of such a choice of the calculated capacity step and feasibility of setting such a limitation on the amount of distribution series members to be determined, which ensure a required accuracy of calculations. The corresponding computational procedure is of sufficient speed. Besides, the unnecessary preliminary equivalenting of the set of units decreases initial data preparation time. In practical calculations, if the amount of groups of units at the node exceeds the limitation specified in the program, then it is reasonable to equivalent single-type units of low capacity.

3.2.6 Multiplication of the Distribution Series of Generating Capacity States at Nodes

Realization of this algorithm in the "KORALL" and "YANTAR" programs is universal. The algorithms allow multiplication of the specified distribution series of states of the generating capacity part of the m-th node (see "generating equipment" in Sect. 3.2.1) and the distribution series of the other part of generating capacity of the m-th node that is calculated in the programs based on the specified equipment set and its performances (see Sect. 3.2.5). In addition, they make it possible to multiply the distribution series of generating capacity states of the m-th node and the energy reliability characteristics (ERChs) of the external transmission lines adjacent to this node.

3.2.7 Calculation of the Distribution Series of States of Transmission Lines in a Separate Tie

Calculation of the function (series) of the distribution of states of transmission lines as well as the states of generating capacity is one of the main operations in assessing the reliability of a complex EPS. For this purpose, an analytical method is used.

The designed algorithm (see Appendix A) is sufficiently universal and enables to calculate the distribution series of states of ties consisting practically of any amount of transmission lines operating in parallel and limit the amount of calculated states to the most probable ones. The corresponding computational procedure is of sufficiently high speed.

3.2.8 Multiplication of the Distribution Series of States of Transmission Lines in a Tie

By analogy with multiplication of the distribution series of generating capacity states, this operation is applied to multiply the given the distribution series of states of part of transmission lines in the n-th tie by the series calculated using the "YANTAR" program.

3.2.9 Generation of System States

Here, generating of system states means determination and ranking of its possible states and their corresponding probabilities by the distribution series of generating capacity states and load schedules with considering of their random variations at nodes, as well as states of transmission lines in ties that are obtained in the program. In this case, the indicated distribution series of system equipment states and the sets of typical daily load schedules for the specified calculated interval are the initial information for the block for determination of the calculated system states. Distribution series and the sets of loads are presented in the ordered form.

To determine the indicators of the reliability of the functioning of the system, it is necessary to analyze only the states characterized by the presence of capacity deficits. In this case, in the state composition algorithm, the possibility of maximum elimination of deficit-free states is provided.

As regards elimination of all deficit-free states, this operation is impossible a priori (without or before the calculation of corresponding states). The problem of elimination of their majority is solved using the expert-logical analysis of the current results on minimization of the calculated deficit states.

In view of the foregoing, the following sequence of calculations for each of S intervals is suggested.

The main idea of the algorithm consists in cyclical enumeration of all possible system states in such a way that each subsequent state would be easier in terms of shortage. The complete enumeration includes three cycles:

- external: enumeration of all possible values of conditional average loads of the ordered schedule for all nodes;
- medium: enumeration of all possible combinations of random load values for the nodes;
- internal: enumeration of all possible combinations of generating capacity states by node and transfer capabilities by ties, which are assigned by the distribution series.

The block of capacity deficit minimization in the calculated states (the CCD block, see A.5) is the core of the internal cycle. The resulting information of this block is used for calculating the reliability indices and for "screening" the deficit-free states. The system states are calculated successively by one, and each next calculated system state is generated after analysis of the previous state in the block of capacity deficit minimization. The decision about the next state to be calculated is made depending on the results of analysis concerning the deficit availability or unavailability in the system.

The developed procedure of screening the deficit-free system states (the algorithm is described in detail in Appendix A) substantially decreases the time for calculation of power system reliability indices.

3.2.10 Minimization of Capacity Deficit (The CCD Block)

The block is intended for nodal capacity deficit minimization based on the specified criterion in terms of provision of the admissible and feasible states and calculation of capacity flows in the network (to determine energy reliability characteristics of transmission lines).

This problem can be solved by several constructed models [9]:

- The two-stage model for capacity deficit minimization. The minimum total deficit in the system is determined at the first stage, and its redistribution among the subsystems—at the second stage. This particular model is realized in the "KORALL" program.
- The model for capacity deficit minimization subject to capacity losses in transmission lines, which are expressed by the quadratic dependence on capacity flows (the "YANTAR" program).
- The model for optimization of calculated states subject to capacity losses and economic characteristics of power generation and transmission ("YANTAR-1" program).
- The model for capacity deficit minimization subject to power losses and electricity markets ("YANTAR-2" program).

The interior point method or its modifications are applied in all the models for capacity deficit minimization [10–20]. The algorithm of this method is not described in the work.

The Model for Calculation of States with Distribution of Capacity Deficit Proportionally to Nodal Loads and with Consideration of Losses Caused by Power Undersupply to Consumers (The CCD Block in "KORALL"—Model 1)

The problem of capacity deficit minimization is solved in two stages and formulated as follows [16]:

> Determine the optimal value of capacity deficit in the system and distribute it optimally by nodes proportionally to loads with regard to the category of consumers and satisfaction of limitations on the ranges of possible variations in generating capacity and the node loads, capacity flows in the ties, as well as maintenance of the balances at the nodes for the known values of serviceable generating capacities, the required levels of load to be served and its significance at the nodes and the specified transfer capabilities of the ties.

The total capacity deficit in the system is determined and minimized at the first stage of the optimization problem solution; i.e., the problem of linear programming with linear constraints is solved.

At the second stage, the calculated capacity deficit is redistributed (if the transfer capabilities of transmission lines allow this) proportionally to the loads at all nodes, provided the determined value of the minimum total deficit in the system does not change; i.e., the nonlinear (quadratic) problem with the same constraints is solved. For the nodes, which are objectively in equal conditions, this principle corresponds to the principle of supporting the equal reliability of power supply to consumers of nodes. In contrast to some other known problems (see Chap. 1), this one has a unique solution and practical application.

When using the interior point method, it is necessary to take into consideration its specific feature that the initial approximation should strictly be a relative interior feasible point of the region determined by the problem constraints in [16–19]. In [17], it is shown that the problem of the second stage has an optimal solution.

Mathematical statement of the problem [9]. The power system calculation scheme is specified as M nodes and N ties between them. The available generating capacity $\overline{P}^{g}_{m}$ and the load $\overline{P}^{l}_{m}\left(m = \overline{1, M}\right)$ are assigned for each node and the maximum feasible bilateral constraints on the transfer capabilities α_n and $\beta_n\left(n = \overline{1, N}\right)$—for each tie, which are different for forward and inverse directions in the general case.

Let us denote the power shortage at the m-th node by y_m, the excess generating capacity by z_m, and the capacity flow in the n-th tie by x_n. Then, the problem of the first stage will be written as follows. Find

$$\min \sum_{m=1}^{M} f_m y_m \tag{3.1}$$

subject to:

$$\sum_{n=1}^{N} \mathrm{a}_{mn} x_n \cdot \quad y_m + z_m = b_m, \quad m = \overline{1, M}; \tag{3.2}$$

$$0 \le y_m \le \overline{P}^{l}_{m}; \quad m = \overline{1, M}; \tag{3.3}$$

$$0 \le z_m \le \overline{P}^{g}_{m}; \quad m = \overline{1, M}; \tag{3.4}$$

$$\alpha_n \le x_n \le \beta_n, \quad n = \overline{1, N} \tag{3.5}$$

where f_m is the objective function coefficient meaning the loss caused by capacity deficit at the m-th node; $b_m = \overline{P}^{g}_{m} - \overline{P}^{l}_{m}\, m = \overline{1, M}$;

a_{mn} are the incidence matrix elements, which take the values:

$$\mathrm{a}_{mn} = \begin{cases} -1, & \text{if the node } m \text{ is the beginning of the tie } n \\ +1, & \text{if the node } m \text{ is the end of the tie } n \\ 0, & \text{if the node } m \text{ does not belong to the tie } n \end{cases} \tag{3.6}$$

Constraint (3.2) is called a "balance equation" and takes into account the first Kirchhoff law, whose sense is "the algebraic sum of capacities flowing to any system node is equal to zero" [1].

At the second stage of problem solution, the principle of proportional distribution of the total capacity deficit in the system among the nodes in terms of the constraints on transfer capabilities of ties is applied. The totality of the capacity deficit $y^{1}_{m}\left(m = \overline{1, M}\right)$ at the nodes, which is obtained at the first stage, is taken as an initial approximation for the interior point method, which is also applied at the second stage. The problem is formulated as the problem of quadratic programming. Find

$$\min\left\{\sum_{m=1}^{M}\left(\overline{y}_m - y_m\right)^2\right\}, \tag{3.7}$$

where $\overline{y}_m = \gamma\,\overline{P}^1_m$ subject to Constraints (3.2)–(3.5).

Here,

$$\gamma = \sum_{m=1}^{M} y^1_m / \overline{P}^1_m \tag{3.8}$$

is the specific value of capacity deficit in the system (kW of deficit/kW of load).

This minimization model for capacity deficit with their subsequent distribution proportionally to nodal loads unambiguously determines both the system and nodal reliability indices, which naturally enhances capabilities and application spheres of computing programs for reliability assessment.

It should be noted that evolution of the above model in the suggested below statement [19] revealed the possibility for improving the results (more rapid convergence of the algorithm):

Find

$$\min \sum_{m=1}^{M} y^2_m / \vec{P}^1_m \tag{3.9}$$

subject to Constraints (3.2), (3.4), and (3.5). Inequality Constraints (3.3) for y_m are not present explicitly (but are taken into consideration automatically in the objective function).

The experimental calculations were carried out for the schemes of very different configurations and with a wide scatter of absolute capacity values (from 10^{-1} to 10^5), which influences the calculation accuracy from the viewpoint of accumulation of calculation errors because of machine representation of numbers and correspondingly the optimization process convergence.

The calculation test scheme is presented in Fig. 3.3.

The main technical and economic characteristics of nodes and ties are shown in Table 3.1. Note that part of the nodes in the scheme is excessive and another part—deficient in terms of the capacity balances (b_m).

The calculation results are compiled in Table 3.2.

The tables show that the excess power (97 MW) in the system (node 6) that was obtained as a result of solution is explained by insufficient transfer capability of the corresponding tie between nodes 5 and 6. Therefore, the system deficit increased by the value of this locked power.

Model for Calculating System States Taking into Account Capacity Losses in Networks (Model 2)

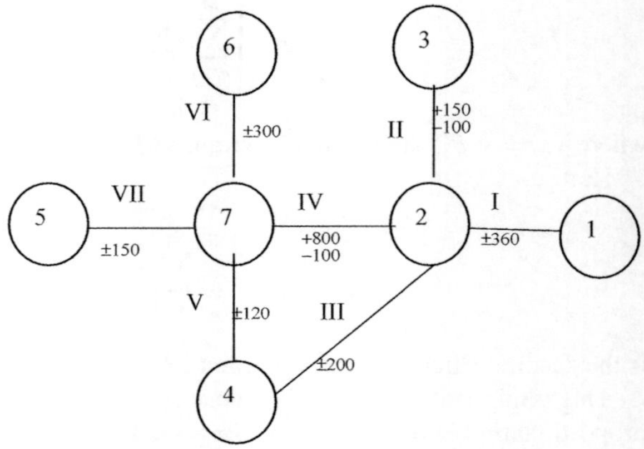

Fig. 3.3 Calculation test scheme

Table 3.1 Characteristic of the calculated operating condition of the test scheme

Nodes of calculation scheme	Generating capacity $\overline{P}^{g}_{m}$, MW	Load power $\overline{P}^{l}_{m}$, MW	Imbalance b_m, MW	Number of tie	Numbers of nodes adjacent to corresponding tie	Transfer capabilities of ties in directions, MW	
						Direct	Inverse
1	2333	2842	−509	I	1–2	360	360
2	1775	1726	49	II	2–3	150	100
3	333	562	−229	III	2–4	200	200
4	1350	170	1180	IV	2–5	800	1000
5	509	1549	−1040	V	4–5	1200	1200
6	921	524	397	VI	5–6	300	300
7	0	196	−196	VII	5–7	150	150
System	7221	7569	−348				

The model is an evolution of the previous model and characterized by the following specific features and advantages (additional capabilities):

(1) indication of an essential factor of EPS operation (costs on capacity transmission), which makes it possible to obtain solutions closer to the factual conditions and physical and technical characteristics of the state and to take into account the second Kirchhoff law roughly;
(2) consideration of capacity losses in networks that is an essential factor for operation conditions of power systems in Russia with sufficiently long transmission lines connecting the nodes;
(3) mathematically realizes one of the possible "dispatcher policies" for setting modes and limiting consumers in the system failure states, providing thereby the uniqueness of the solution;

Table 3.2 Calculation results of the test scheme by Model 1

Nodes of calculation scheme	Capacity deficit y_m, Stage I, MW	Capacity deficit y_m, Stage II, MW	Excess capacity z_m, MW	Number of tie	Numbers of nodes adjacent to tie	Calculated capacity flows in ties in directions x_n, MW	
						Direct	Inverse
1	203.1	157.5	0	I	1–2	–	351.6
2	44.6	86.9	0	II	2–3	150	–
3	91.5	79.0	0	III	2–4	–	104
4	13.2	0	0	IV	2–5	–	261
5	43.9	75.5	0	V	4–5	1076	–
6	0	0	**97**	VI	5–6	–	300
7	48.6	46.0	0	VII	5–7	150	–
System	444.9	444.9	97				

(4) replacement of the two-stage solution of the capacity deficit minimization problem with the single-stage one, which substantially decreases the problem solution time.

In the statement of the problem, in addition to the above-mentioned initial information, the new is the accounting of power losses in the transmission lines using the "loss factor" $k_{\text{loss}\,n}$, which is determined from the values of the parameters r_{0n}, l_n, P_n, U_n, and $\cos\varphi_n$ (see below). Calculation of losses is carried out on the value of net capacity flow P_n tie.

Mathematical formulation of the problem.

Leaving the same notation for variables, the problem of determining the total capacity deficit of the system, taking into account the losses in the communication lines, is formulated as follows:

$$\text{Find } \min \sum_{m=1}^{M} f_m y_m \tag{3.10}$$

subject to

$$\sum_{n=1}^{N} a_{mn} x_n + \sum_{n \in J_m^*} b_{mn} k_{\text{loss}\,n} x_n^2 - y_m + z_m = l_m, \quad m = \overline{1, M}, n = \overline{1, N} \tag{3.11}$$

$$0 \le y_m \le P_m^{1} \quad m = \overline{1, M}, \tag{3.12}$$

$$0 \le z_m \le P_m^{g} \quad m = \overline{1, M}, \tag{3.13}$$

$$\alpha_n \leq x_n \leq \beta_n \quad n = \overline{1, N}. \tag{3.14}$$

Here,

The matrix $\{a_{mn}\}$ was determined in (3.6).
The set J_m^* represents the nodes receiving capacity flow over the ties, i.e.,

$$J_m^* = \{m : a_{mn} x_n > 0\} \quad m = \overline{1, M}, \; n = \overline{1, N}; \tag{3.15}$$

The matrix $b_{mn} = \begin{cases} 1, \text{ if } a_{mn} \cdot x_n > 0 \\ 0, \text{ if } a_{mn} \cdot x_n \leq 0 \end{cases}$ takes account of losses in the tie of the receiving node and in essence represents the set J_m^*.

The point with $x_n = 0$ is taken as a starting point in the block of searching for the initial approximation by the independent variables $y_m, z_m, \left(m = \overline{1, M}\right)$, $x_n, \left(n = \overline{1, N}\right)$ are chosen from balance Eq. (3.11) in terms of the remaining Constraints (3.12)–(3.15). Deviation from the admissible region boundaries for y_m and z_m is taken at a level of 10% of the corresponding boundary value.

The dual variables are analyzed to reveal nodes with capacity deficit and non-deficit.

When using the interior point method, the matrices $\{a_{mn}\}$ and $\{b_{mn}\}$ are not calculated and stored, since they consist of zeros and ± 1.

All calculations are carried out using the identification numbers of the nodes of the beginnings and the ends of the ties.

The formulated problem is a typical nonlinear programming (NP) problem. The objective function is linear, equality Constraints (3.11) are nonlinear, and Constraints (3.12)–(3.15) are given as two-sided inequalities [17, 20–23].

As already noted, accounting for losses in lines approximately "replaces" the accounting of the second Kirchhoff law and leads to an unambiguous solution of the problem. This was confirmed by comparing the flow distribution calculations performed according to the described algorithm and the program of electrical calculations of SDS-5 [24]. In most comparative calculations, the differences in active power fluxes did not exceed 1–5%. Moreover, it was possible to reach practically complete coincidence by selecting the loss factors.

The model taking into account in transmission lines losses can find application in the practice of operational control of EPS as in the preparation of planned modes with the expected capacity deficit and as an adviser to the dispatcher in the current control process. Thereby, the accuracy is assured by the proper selection of the loss factor (not by the average one, but real values of U_n and $\cos \varphi_n$).

The suggested algorithm can contribute to implementation of the following control concepts (depending on the expected level of deficit, the place of its source, and the system structure):

- accounting of the consumer category by setting the corresponding weighting coefficients f_m in different forms (in the form of specific damages, penalties, factor of load significance, etc.);

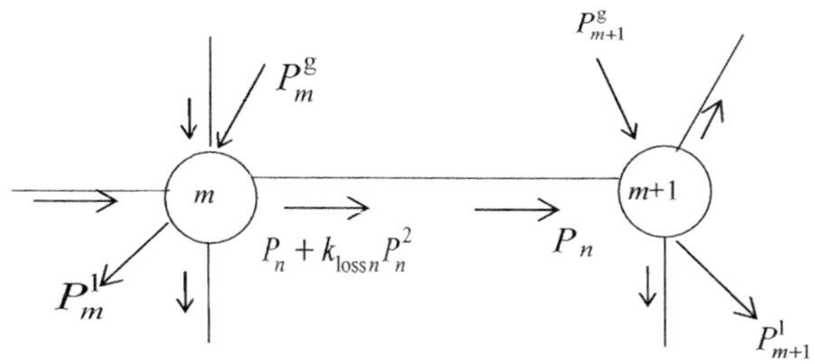

Fig. 3.4 Illustration of accounting of network losses

- accounting of network losses by setting the corresponding values of $k_{loss\,n}$ in parallel with the consumer category or without it;
- "elimination" of deficit by reducing the frequency. In this case, it is assumed that $f_m = 1$, $m = \overline{1, M}$, and $k_{loss\,n} = 0$, $n = \overline{1, N}$, and the solution of y^1_m obtained at the first stage by the mentioned program is supplemented by the second stage with minimization of the functional

$$\sum_{m=1}^{M}\left(\left(\sum_{m=1}^{M} y^1_m\right)\frac{K_m^{slope}}{K_{syst}^{slope}}\cdot\frac{P_m^{slope}}{\sum_{m=1}^{M} P^l_m} - y_m\right)^2 \to \min,$$

where K_m^{slope}, K_{syst}^{slope} are the coefficients of the frequency curve slope of the m-th node and the system.

The application of the proposed computational tool is most effective at the upper hierarchical levels (UPS, IPS), for which the computational schemes are usually presented so enlarged that it makes no sense to use models of accurate electrical calculations. However, in this case, the losses in external ties turn out to be of the same order with the disconnected load.

Capacity (electric power) losses in networks play an essential role in optimization of deficient postemergency states (modes), because the vast majority of capacity imbalances do not exceed 10–20% of consumption and the losses are of the same order (5–15% of consumption). The real value of losses in UPS of Russia in 1995, for example, was 10.3% of the power production. This fact makes possible consideration of network losses necessary to enhance the model quality.

An essential feature of taking into account the losses in the models of calculation of the states (modes) of the system is also the fact that this makes it possible to obtain a single and feasible solution, which is often impossible in the case of a linear model for optimizing capacity deficits in the system.

The accounting principle of capacity losses in transmission lines is explained in Fig. 3.4, where $k_{loss\,n}$ is the loss factor in the n-th transmission line and the losses are calculated using the below formulas, which are certainly approximate, but, nevertheless, acceptable for the considered case.

The capacity loss factor is determined based on the formula of power losses in the transmission line

$$P_{\text{loss}\,n} = 3 \cdot r_n \cdot I_n^2, \tag{3.16}$$

where r_n is the active resistance of one phase of the transmission line,

I_n is the total current in the transmission line.

Provided that $r_n = r_{0\,n} \cdot l_n$, and $I_n \quad \frac{P_n}{\sqrt{3}U_{n\,\text{cp}} \quad \cos\varphi_{n\,\text{cp}}}$, Formula (3.16) can be written as follows $P_{\text{loss}\,n} \cong \frac{3 \cdot r_{0\,n} \cdot l_n}{3 \cdot U_{n\,\text{average}}^2 \cdot \cos\varphi_{n\,\text{average}}^2} \cdot P_n^2 = k_{\text{loss}\,n} \cdot P_n^2$,

where

$$k_{\text{ПОТ}\,n} = \frac{r_{0n} \cdot l_n}{U_{n\,\text{cp}}^2 \cdot \cos\varphi_{n\,\text{cp}}^2}, \tag{3.17}$$

$r_{0\,n}$	is the specific active resistance of the n-th transmission line, ohm/km;
l_n	is the transmission line length, km;
$U_{n\,\text{average}}$	is the average voltage value of the n-th transmission line, kV;
$\cos\,\varphi_{n\,\text{average}}$	is the average value of the capacity factor of the n-th transmission line.

Determination of the loss factor $k_{\text{loss}\,n}^{\text{tie}}$ for the tie as a set of currents of I_n transmission lines of different lengths, voltages, and $\cos\varphi x$ is a sufficiently complicated problem. For the solved optimization problem of calculated states (mode), the loss factor is determined as:

$$k_{\text{loss}\,n}^{\text{tie}} = \sum_{i_n}^{I_n} k_{\text{loss}\,i_n} \cdot \alpha_{i_n}^2,$$

where $\alpha_{i_n} = \frac{P_{i_n}}{\sum_{i_n}^{I_n} P_{i_n}}$ is the loading level of each transmission line in the tie or $\alpha_{i_n} = \frac{\overline{P}_{i_n}}{\sum_{i_n}^{I_n} \overline{P}_{i_n}}$, where $\overline{P}_{i_n}$ is the transfer capability of the i_n-th transmission line in the n-th tie.

For the operating control, the loss factors can be calculated based on the actual data on network functioning.

Comparative analysis was carried out using the calculation scheme in Fig. 3.3 with the same initial information. Table 3.3 presents more detailed initial information for the ties.

The calculation results (see Table 3.4) vividly show that if the capacity losses in networks are taken into account, the total capacity deficit in the system increases roughly by the value of losses in comparison with the previous model (see Table 3.2).

This fact is natural, since the coefficients f_m in the minimized functional in this calculation are taken equal to 1; i.e., the nodal loads are equivalent. In such a case, if the losses in transmission lines are neglected, the available generating capacities will cover a greater volume of load.

This result (the shortage growth approximately by 4%) is quite expected, which confirms correctness of the program work and problem statement.

Table 3.3 Characteristic of interconnection ties of the test scheme

Number of tie (connected nodes)	Number of transmission lines in tie	Transfer capabilities of transmission lines in directions, MW		Transmission line length, km	Voltage, kV	Loss factor k_{loss}, 10^{-3}
		Direct	Inverse			
I (1–2)	1	180	180	183	220	0.14
	2	180	180	223	220	0.17
II (2–3)	1	75	50	250	220	0.18
	2	75	50	250	220	0.18
III (2–4)	1	200	200	300	220	0.31
IV (2–5)	1	150	150	410	220	0.42
	2	150	150	95	220	0.10
	3	500	700	580	500	0.03
V (4–5)	1	900	900	360	500	0.02
	2	150	150	120	220	0.12
	3	150	150	187	220	0.19
VI (5–6)	1	150	150	185	220	0.19
	2	150	150	185	220	0.19
VII (5–7)	1	75	75	830	220	0.85
	2	75	75	830	220	0.85

Table 3.4 Calculation results by Model 2 of the calculation test scheme

Nodes of calculation scheme	Capacity deficit, MW	Excess capacity, MW	Number of tie	Numbers of nodes adjacent to tie	Capacity flows in directions, MW		Capacity losses in ties, MW
					Direct	Inverse	
1	324.6	0.0	I	1–2	0	184.5	2.65
2	1.9	0.0	II	2–3	146.4	0	1.07
3	82.6	0.0	III	2–4	0	188.6	1.64
4	0.498	0.0	IV	2–5	0	95.1	0.15
5	1.1	0.0	V	4–5	982.1	0	8.68
6	0.0	96.39	VI	5–6	0	299.9	0.72
7	47.8	0.0	VII	5–7	148.19	0	0.20
System	459.7	96.39	Total losses				15.01

Distinctly different distribution of capacity flows in the ties, and capacity deficit at the nodes is obtained using this model in comparison with the previous one. This is explained by the fact that in the previous model the power flows were distributed in proportion to the nodal loads. The solution obtained using this model is more realistic (natural).

Model of Optimization of Calculated States (Modes) of the System Taking into Account Economic Characteristics of Generation and Transmission of Electric Power (Model 3)

The processes going on in the Russian economy greatly affected electric power industry that switched from the purely centralized model of management to the model with elements and criteria of market relations and mechanisms. Their introduction in electric power industry, conversion of power systems into joint-stock companies, arising competition all placed new demands on the research methods including the reliability research and correspondingly on the problems of optimization calculated states (modes) system and capacity deficit minimization [24].

The strategy of power system behavior in emergency and postemergency states (modes) caused by suddenly occurring considerable generating capacity deficit and/or insufficient transfer capabilities of ties should include fast calculated states adjustment, which is coordinated with all members of the interconnection which contribute, if possible, to the least losses because of interruption of power supply to consumers, as well as the minimum financial expenses for power generation and distribution in power systems.

The analysis of deficit-free states (modes) is necessary in the reliability models taking into account provision with primary energy resources, which became topical in the formed situation. It is natural that in the deficit-free state, the functional of type $\min \sum_{m=1}^{M} f_m \cdot y_m$ cannot be an optimization criterion, since it degenerates into zero. For optimization of any calculated states, the functional is supplemented with the criterion of power production cost at nodes and the expenditures for its transmission in ties. The rest of requirements and constraints do not change compared to the previous model. Besides, the obtained calculated states (modes) should be admissible and feasible in terms of physical and technical characteristics of the power system. In the market economy environment, the extreme strategy is the requirement that each power system experiencing an emergency should make efforts for eliminating its consequences first of all by itself including interruption of power supply to its consumers in the required volume. The degree of emergency mutual assistance depends on the interconnection structure, the parameters of power systems, and interconnection ties between them, as well as on the commercial effectiveness of such assistance for the interconnection members.

It should be noted that only the effectiveness calculations with reliability accounting can show technical and economic expediency of different strategies for power system development and establish ratios among different types of effectiveness, which take account of the interests of the stakeholders and their individual subjects. These calculations can be made on the model for calculations of normal and postemergency, both deficit and deficit-free, states.

The criteria, which determine the effectiveness of power system control in the suggested model, involve electricity rates, damage caused by power undersupply to consumers, fixed costs on its production, and also variable costs characterizing primarily the cost of consumed primary energy resources.

In the indicated statement, the problem of deficit optimization is formulated as follows:

> Determine the optimal loading of generating equipment by node, the optimal values of the served load at nodes based on the specified optimality criterion for the known values of available generating capacity, specific fixed and variable costs on power production, required

levels of load to be served, electricity rates and compensation expenditures for unserved load at nodes, as well as specified transfer capabilities of ties and power loss factors in them provided that the limitations on the ranges of possible changes in generating capacity and loads at nodes, power flows in ties are satisfied and the nodal power balances are maintained.

Mathematical formulation of the problem. The optimization problem of any (deficit and deficit-free) system state can be formulated in the following way. Find

$$\max \sum_{m=1}^{M} \left\{ c_m \left(\overline{P}^{1}_{m} - y_m \right) - f_m y_m - d_m \left(\overline{P}^{g}_{m} - z_m \right) + f l_m z_m \right\}, \tag{3.18}$$

where c_m is the price (rate) of electricity supplied to consumers at the m-th node (m.u.[1]/kW h); f_m is the specific damage or compensation expenditures due to electricity undersupply (m.u./kW h); d_m is the specific costs on production and transmission of 1 kW h at the corresponding node (m.u.); fl_m is the cost of fuel required for production of 1 kW h (m.u.); and with the constraints completely coinciding with Constraints (3.12)–(3.15) of the previous model. In the general case, the optimal solution can be obtained at real relationships among the functional coefficients $f_m > c_m$, $c_m > d_m$, $d_m > fl_m$ in the majority of system nodes. In addition, it is conventionally assumed in the model that the duration of the considered calculated state is equal to 1 h, and hence, the capacity (kW) numerically coincides with power production (kW h).

This model allows optimization of the states caused not only by an emergency situation in one of power systems, which operates in parallel with others, but by several coinciding emergencies in different parts of the interconnection. This property of the model makes it possible to devise sufficiently universal strategies ensuring high effectiveness of power system control practically for any situations.

The contractual power supplies from system to system can be maintained in the model by fixing the capacity flow in the corresponding ties at the preemergency or reduced (as a function of contract terms) level.

Thus, for the suggested problem statement, an additional model is constructed to optimize both shortage and shortage-free conditions of power systems, and as a result, more complete information, in particular on transmission line loading, is obtained. For the state estimation, the optimization model in such a statement provides economic management of technical resources of power systems in power production and distribution. Thus, this model can be applied to reliability calculation in terms of economic characteristics of system operation.

Results of the comparative analysis. The researches were conducted on the schemes of most different configurations. Here, the results of comparison with the previous models are presented for the scheme in Fig. 3.3.

The cost indices presented in Table 3.5 are the additional information, and the calculation results are shown in Tables 3.6 and 3.7.

---

[1] m.u.—monetary units.

Table 3.5 Cost indices for the test scheme (Fig. 3.3)

Nodes of calculation scheme	Electricity price c_m, m.u./kW h	Compensation expenditures (losses) f_m, m.u./kW h	Full costs on electricity production d_m, m.u./kW h	Cost of fuel for electricity production fl_m, m.u./kW h
1	0.036	0.75	0.032	0.015
2	0.063	0.75	0.030	0.010
3	0.048	0.75	0.040	0.015
4	0.030	0.50	0.010	0.001
5	0.048	1.00	0.030	0.015
6	0.038	0.75	0.030	0.015
7	0.048	1.00	0.040	0.020

Table 3.6 Calculation results of the test scheme

Nodes of calculation scheme	Capacity deficit, MW	Cost of electricity supplied to consumer, thousand m.u.	Loss caused by electricity undersup-ply, thousand m.u.	Costs on electricity production by node, thousand m.u.	Cost-effectiveness, thousand m.u.	Financial effective-ness, thousand m.u.
1	172.8	83.99	129.58	74.66	1927.3	9.33
2	0.8	111.83	0.6	53.25	1240.7	58.6
3	80.4	15.98	60.3	13.32	347.9	2.7
4	169.7	40.50	84.8	13.50	– 13.4	27.0
5	0	24.43	0	15.27	1533.7	9.2
6	0	31.33	0	26.18	366.8	5.2
7	46.1	0	46.2	0	149.8	0
System	469.8	308.06	321.5	196.18	5552.6	111.9

Table 3.7 Resulting information on tie loading

Number of tie	Calculated capacity flows (MW) in ties in directions		Capacity losses, MW
	Direct	Inverse	
I	–	336.3	8.82
II	148.7	–	1.11
III	–	193.9	1.73
IV	–	251.2	1.07
V	1142.3	–	11.74
VI	–	299.9	0.72
VII	–	149.8	0.20
Total losses			25.39

Summary of the suggested model:

- The model of optimization of deficit and deficit-free states (modes) of EPS is formulated and implemented for its use in reliability assessment programs. Evaluation in the optimization block of various calculated states (modes) allows to take into account in the reliability assessment program the provision of primary energy resources and to estimate the loading of ties during the calculation period.
- The performed calculations of the efficiency of various calculation states (modes) of energy associations prove the possibility of quantitative justification of various management models and measures to ensure reliability in a competitive environment. Research using this model will make it possible to formulate a set of rules for the joint work of private energy companies.
- The model can also be used in the dispatch centers of power companies for making decisions on maintenance of high reliability and cost-effectiveness of power supply to consumers and also for research purposes.

Model for Optimization of Calculated States (Modes) of the System Taking into Account Characteristics of the Wholesale Electricity Markets (Model 4)

Change in the social and economic conditions of power system development in Russia called for adaptation of the formed methodology of power system development management, change and supplement of the set of applied mathematical models and methods, certain corrections of the known mathematical models and programs. For the liberalized electricity and capacity market, there is a lack of vertical integration and only horizontal integration and, as a result, increased market competition between both producers and energy suppliers. In addition, priority tasks have come to the fore, the solution of which must be taken into account in reliability assessment models. Among them:

1. Providing a more complete and methodical sound accounting for the supply of primary energy resources to all types of power plants, and primarily thermal ones, operating on organic fuel. But in this case, an important role is played by the accounting of fluctuations in the inflow of water into the reservoirs of the hydroelectric power station. In the planned economy, the fuel was supplied to power plants in the first turn regardless of its price and transportation expenses, and this was the state privilege. In the market environment, the fuel to power plants is delivered on a contractual basis. Sometimes, the fuel transportation expenses are overestimated, and fuel supplies are disrupted because of the crisis phenomena in the society and the national economy, etc.
2. Revision of the approaches to assessment of the reservation levels in power systems, substantiation of the necessity, and the rational volume of the so-called commercial (market) generating capacity reserve. The market is known to demand additional expenses for reserves by virtue of the higher uncertainty of power system development in comparison with the planned economic management. Additional redundancy is necessary for ensuring competitiveness of power companies and for earning additional profit. The higher reservation can be caused by

the legislative and contractual obligations of power companies before consumers for power supply reliability.
3. Accounting of disintegrating processes arising as a result of extensive conversion of power systems into joint-stock companies and occurrence of small power companies in the power system reliability models. This fact demands first of all the change of the power system reliability support principle “top-down” to the principle “down-top.”
4. Accounting of occurrence and functioning of the wholesale markets of different levels, opportunities of these markets in the case of power shortages in the system.
5. Accounting in the unit of optimization of the calculation states (modes) of EPS during the investigation of their reliability of various ways of mutual assistance of electric power companies to each other, depending on the contractual relations between them.

For the most part, the indicated problems are not novel for the western power companies. And hence, to research the reliability of EPS in foreign models, these or other solutions were obtained. This to some extent facilitates the work of Russian specialists in the modernization of their programs. But, on the other hand, the historical development of Russia’s EPS has its own specific features, which should be taken into account in the models by all means.

In the current context, the presented necessary changes in the methods and models for power system reliability study alter the computational algorithm of reliability assessment rather markedly. This concerns primarily a set of calculated system states to be optimized in the power system reliability analysis. In previous models, this set consisted only of postemergency deficient states (regimes). Now, in order to take into account the fuel costs and the impact on the functioning of wholesale markets, it is necessary to consider all possible states (regimes)—of course, statistically. Consideration of all calculated states of EPS (deficit and deficit-free) offers additional possibilities for evaluating of distribution functions of transmission line loading. These distribution functions are the information required for the choice of the transfer capability of ties between the nodes [25–27].

The restructuring of electric power industry management in Russia and its regions is going on with sufficient difficulties, not always reasonably and rationally. For example, the structure of the wholesale electricity and capacity market has not taken shape definitely, and the issues of specific features of the Russian economy have not been investigated to the full.

New interrelations between electricity producers and consumers, as well as the happened administrative and economic separation of the systems within the former UPS, did not decrease the number of problems in reliability research. On the contrary, they gave rise to some new problems indicating that reliability, in particular the system reliability in power supply to consumers, becomes highly topical. This, in turn, necessitates expansion of the calculation schemes in the number of nodes and ties between them.

It should also be mentioned that consideration of the limitations on primary energy resources for reliability assessment involves a choice of the strategy and tactics of their use, which should be presented in the model algorithm. There are no well-defined principles of such a choice. In the actual practice, they are met in essential diversity. Moreover, this diversity is explained both by the objective reasons (such as a real share of HPPs in the system and high uncertainty of forecasting the subsequent conditions of power system operation), and the subjective factors (qualification of the personnel, value orientations of the managers of the systems and rules of the regional energy commission, short-term benefits and methods for limiting consumers: frequency and voltage reduction, "littered" restrictions, fan (rotational) shutdowns, etc.). Therefore, the problem of considering diversity and variety of the principles of choosing the strategy and tactics in system operation modeling still remains unsolved.

The commercial reserve size should be estimated quantitatively to decrease the loss caused by the possible commercial risk $\boldsymbol{r}$ of power undersupply to consumers.

In [25], it is showed that the commercial capacity reserve in a simplified form can be equal to:

$$R^{\mathrm{el}}_{\mathrm{commer}} = p_{\mathrm{r}}\mathrm{E}_{\mathrm{dem}},$$

where p_{r} is the rate of return against the production cost, $\mathrm{E}_{\mathrm{dem}}$ is the electricity demand, and the commercial capacity reserve is

$$R^{\mathrm{capacity}}_{\mathrm{commer}} = R^{\mathrm{el}}_{\mathrm{commer}}/T_{\mathrm{instal}},$$

where T_{instal} is the installed capacity utilization time of generating sources (h).

The need to take account of the division of national power systems into financially independent subsystems, while keeping the energy and electric unity of operating states (modes) of all systems, expands the list of optimization criteria for power system states in terms of the economy, including:

- the minimum fuel consumption or the fuel costs;
- the minimum operating costs;
- the maximum profit;
- the maximum or minimum price in the market and others.

In the case of centralized management, the chosen criterion should ensure the effective operation of UPS within the whole national economy. The financial interests of individual power companies (subsystems) often contradict this criterion. Therefore, the models have to include the possibility for coordination of contradictory or heterogeneous criteria with the help of the corresponding software, which takes into account decentralization. In the market economy, these contradictions are analyzed by supporting the balance of demand and supply in the monetary terms.

Practice [28] of different countries and power companies shows that the principles of arrangement of the wholesale electricity and capacity markets can differ. The main difference involves formation of electricity prices (rates) as a function of the wholesale market level; i.e., there exist internal rates, which can be different (rates of power companies and rates of the regional wholesale market of power pools and the federal wholesale market within the unified power system of the country).

The ratio of the wholesale electricity prices should be based on the conformity principle of the tasks aimed at fuel saving and rational electricity consumption.

In the suggested model, the strategy of the wholesale price formation can be established by several methods. Although, in general, any strategy can be ignored by setting the specific cost of purchased and sold electricity for nodes.

Nevertheless, we will describe two possible strategies.

Strategy I is proposed in [27]:

- The generating companies (GCs) make applications for electricity sale through the wholesale market (price, volume).
- The consumers make applications for electricity purchase through the wholesale market (volume).
- The wholesale price is chosen starting with the minimum price suggested in the applications. All electricity prices and its corresponding volume participating in provision of the subscribed load are analyzed in an ascending order. The choice terminates on the price of the GC, which completes the coverage of the load on the system. Thus, the remaining GCs do not participate in the wholesale market at the very moment and hence suffer that they excessively overpriced and cannot sell their excesses. And thus, the wholesale price is defined as the maximum, participating in the coverage of the load.

Strategy II

The price in the wholesale market is attached to each GC according to its application (it can be equal for all of them as well).

The previous optimization model in the assessment of EPS states allowed for the economic management of the system's technical resources in the production and distribution of electric capacity. Model 4 described below is a development of model 3. In addition to the technical and economic characteristics of the production and transmission of electricity, features of the regional and federal energy markets are taken into account. This since these features affect the results of the distribution of deficit of capacity and optimization of deficit states and deficit-free states (mode) [26].

In this context, the problem of calculated states (modes) optimization can be formulated as follows:

> Determine loading of the generating equipment and the value of the load to be covered in the nodes according to the specified optimality criterion, taking into account the limitations on the ranges of possible changes generating capacity and a load at nodes, capacity flows of ties, and also of perform balances the nodes, taking into account the power loss in the networks for the known values of the available generator power, for the specific costs for generating electricity, for the required coverage levels and the importance factors for the loads in the nodes, for the given communication capacities and for the capacity loss factors therein, as well as for the tariffs for electricity for the domestic market of each power center and external wholesale markets.

Mathematical formulation of the problem. The problem functional has the form:

$$\max \sum_{m=1}^{M} \left\{ c_m \left(\overline{P}_m^{1} - y_m \right) + c_m^{el} \sum_{n \in N_0} P_n - f_m y_m - d_m \left(\overline{P}_m^{g} - z_m \right) + f l_m z_m \right\}, \tag{3.19}$$

where N_0 is a set of ties, via which the m-th node sales (purchases) electricity in the wholesale market.

The problem constraints remained the same as in the model that took into account only power losses [see (3.11)–(3.15)].

In the functional (3.19) of model 4, in contrast to the previous model 3, there is a coefficient c_m^{el} that is the electricity price offered by the m-th node in the wholesale market (m.u./kW h), and there is a small choice of possibilities for its assignment:

1. c_m^{el}—as the wholesale (selling) electricity price assigned by the Federal wholesale market for the m-th node.
2. c_m^{el}—as the wholesale (selling) electricity price that is equal for all nodes and chosen based on one of the suggested strategies of the wholesale market arrangement, e.g., in [27].

A feature of recording a functional is that the dimensionality of the measurements is not observed, since specific costs referred to per kilowatt hour are multiplied by the capacity (kW). The lifetime of this mode, common to all components of the functional, is taken out of the brackets and is determined outside the framework of this optimization block in program.

As was already said, Problems (3.19) and (3.11)–(3.15) were also solved by the interior point method, which proved to be rather flexible, of high speed and fast convergence for the similar problems.

In the presented statement, the "YANTAR" program can solve the indicated above problems. In this context, the reliable power supply to consumers was researched depending on provision with primary energy resources and in combination with provision with generating capacities.

The "YANTAR" program is also adjusted for evaluation of the required capacity reserve level in each regional power system for the case of its isolated operation at the given $\mathscr{P}$.

Processing of the calculation results. The initial cost coefficients by node are $c_m, c_m^{el}, d_m, f l_m, f_m$.

The *initial* data characterizing power system operation include:

- operating generating capacity at $P^{g}_{\text{op}\,k_{ms}}$ the node (MW);
- nodal load $\overline{P}_m^{1}$ (MW);
- transfer capabilities P_n of ties in the direct and inverse directions (MW).

The calculated data by node are:

- capacity deficit $\Delta P_m^{\mathrm{l}} = y_m$ for consumers at the nodes (MW);
- excess generating capacity $\Delta P_m^{\mathrm{g}} = z_m$ (MW);
- load provided by coating $P_m^{\mathrm{l}} = \overline{P}_m^{\mathrm{l}} - \Delta P_m^{\mathrm{l}}$ (MW);
- generating capacity used to cover the load $P_m^{\mathrm{g}} = \overline{P}_m^{\mathrm{g}} - \Delta P_m^{\mathrm{g}}$ (MW);
- the sum of the flows of capacity $P_{\mathrm{in}\,m}$ over in ties <u>**to**</u> the m-th node;
- the sum of capacity flows $P_{\mathrm{out}\,m}$ in ties <u>**from**</u> the m-the node.

The resultant economic indicators for nodes and the system are represented in the following way (conditionally attributable to one hour).

1. The cost Z_1 of electricity supplied to consumers that includes (m is the number of the node):

$$Z_{1\,m} = Z_{1\,m}^0 + Z_{2\,m}^0,$$

where

$$Z_{1\,m}^0 = \begin{cases} \left[\overline{P}_m^{\mathrm{l}} - y_m - (P_{\mathrm{in}\,m} - P_{\mathrm{out}\,m})\right] \cdot c_m, & \text{if } P_{\mathrm{in}\,m} > P_{\mathrm{out}\,m} \\ \left(\overline{P}_m^{\mathrm{l}} - y_m\right) \cdot c_m, & \text{if } P_{\mathrm{in}\,m} < P_{\mathrm{out}\,m} \end{cases}$$

is the cost of electricity supplied to own consumers;

$$Z_{2\,m}^0 = \left[\overline{P}_m^{\mathrm{g}} - z_m - \left(\overline{P}_m^{\mathrm{l}} - y_m - (P_{\mathrm{in}\,m} - P_{\mathrm{out}\,m})\right)\right] \cdot c_m^{\mathrm{opt}}$$

is the cost of electricity supplied to external consumers at the wholesale price c_m^{opt} that is correspondingly equal to c_m^{el} or at some other fixed or pre-calculated price.

2. The damage caused by electricity undersupply to consumers

$$Z_{2\,m} = f_m y_m$$

3. The costs on electricity production:

$$Z_{3\,m} = d_m \cdot \overline{P}_m^{\mathrm{g}} - fl_m \cdot z_m,$$

where $fl_m \cdot z_m$ is the saved fuel cost.

4. The national (economic) effect for the system

$$\mathrm{Ef} = \sum_{m=1}^{M} \left[f_m\left(\overline{P}_m^{\mathrm{l}} - y_m\right) - (d_m - fl_m)\overline{P}_m^{\mathrm{g}} - fl_m\left(\overline{P}_m^{\mathrm{g}} - z_m\right)\right],$$

where

$f_m\left(\overline{P}^{\mathrm{l}}_m - y_m\right)$ is the prevented damage.

$fl_m\left(\overline{P}^{\mathrm{g}}_m - z_m\right)$ are the fuel costs.

$(d_m - fl_m)\overline{P}^{\mathrm{g}}_m$ are the fixed costs.

Breaking down the second and third subtrahends in the expression of the economic effect, we obtain the following formula:

$$-(d_m - fl_m)\overline{P}^{\mathrm{g}}_m - fl_m\left(\overline{P}^{\mathrm{g}}_m - z_m\right) = -d_m\overline{P}^{\mathrm{g}}_m + fl_m\overline{P}^{\mathrm{g}}_m - fl_m\overline{P}^{\mathrm{g}}_m + fl_m z_m$$
$$= -d_m\overline{P}^{\mathrm{g}}_m + fl_m z_m.$$

It follows that the economic effect represents some gained effect (let us call it the "prevented damage," Table 3.10) minus the total costs on electricity production in terms of the cost of saved fuel for the unused capacity ($fl_m z_m$). This subtrahend makes it possible to analyze more definitely the situations, when the node prefers to purchase electricity from other nodes (as it is cheaper because of expensiveness of fuel used at the node) rather than produce it.

5. the commercial effect (profit) in the system.

In general, the profit can be represent so:

$$\boxed{\mathrm{Pr} = \sum_{m=1}^{M} (Z_{1m} - Z_{3m})}$$
$$= \boxed{\text{Cost of electricity supplied to consumers}} - \boxed{\text{Electricity production costs}}$$

The use of the generation P^{g}_m to serve load and the served load P^{l}_m rather than power flows in ties enables to calculate the profit as

$$\mathrm{Pr} = \sum_{m=1}^{M} \begin{cases} (c_m - d_m)\cdot P^{\mathrm{l}}_m + \left(c^{\mathrm{opt}}_m - d_m\right)\cdot\left(P^{\mathrm{g}}_m - P^{\mathrm{l}}_m\right), \text{ if } P^{\mathrm{g}}_m \geq P^{\mathrm{l}}_m; \\ (c_m - d_m)\cdot P^{\mathrm{g}}_m + \left(c_m - c^{\mathrm{opt}}_m\right)\cdot\left(P^{\mathrm{l}}_m - P^{\mathrm{g}}_m\right), \text{ if } P^{\mathrm{g}}_m < P^{\mathrm{l}}_m. \end{cases}$$

In this expression, the profit for the system as a whole includes the profits from:

(a) The cover the load of own consumers and part of the electricity supplied to the wholesale market, *if generation at the node exceeds load.* Besides, here P^{g}_m is used generation to cover the load $P^{\mathrm{g}}_m \leq \overline{P}^{\mathrm{g}}_m$, and P^{l}_m is covered part of the load $P^{\mathrm{l}}_m \leq \overline{P}^{\mathrm{l}}_m$;
(b) The cover the load of own consumers that is equal to generating capacity of the m-th node, and the profit expressed by the summand, which indicates the difference between the electricity price c_m for own consumers and the wholesale price c^{el}_m (or c^{opt}_m), i.e., $\left(c_m - c^{\mathrm{opt}}_m\right)\left(P^{\mathrm{l}}_m - P^{\mathrm{g}}_m\right)$. If it turns out that the remaining part of the load $\left(P^{\mathrm{l}}_m - P^{\mathrm{g}}_m\right)$ is to be served by the electricity purchase at a higher

price, i.e., $c_m^{\mathrm{el}} > c_m$, the profit decreases, and if it is to be purchased at a cheaper price, i.e., $c_m^{\mathrm{el}} < c_m$, the profit increases.

The profit Pr can also be determined by other methods, e.g.:

$$\mathrm{Pr} = \sum_{m=1}^{M} \begin{cases} c_m P_m^{\mathrm{l}} + \left(c_m^{\mathrm{el}} - fl_m\right)\left(P_m^{\mathrm{g}} - P_m^{\mathrm{l}}\right) - (d_m - fl_m)\overline{P}_m^{\mathrm{g}}, & \text{if } P_m^{\mathrm{g}} \geq P_m^{\mathrm{l}}; \\ c_m P_m^{\mathrm{l}} + \left(c_m - c_m^{\mathrm{el}}\right)\left(P_m^{\mathrm{l}} - P_m^{\mathrm{g}}\right) - d_m P_m^{\mathrm{g}}, & \text{if } P_m^{\mathrm{g}} < P_m^{\mathrm{l}}. \end{cases}$$

The following rules of the wholesale electricity market performance, which are inherent in this model, can be formulated on the basis of the above said.

1. Each electricity producer m has the right to be a subject of the wholesale market (purchaser or seller).
2. All subjects of the wholesale market enjoy equal opportunities for using any transit transmission lines.
3. The volume of electricity supplied to the wholesale market for selling is established in terms of the load of own consumers and the reserves for repairs.
4. The capacity supplied to the wholesale market and received from the wholesale market loses its targeting, this being most likely a model drawback.

Practical application of the model. This model in the form of a subprogram is also part of the "YANTAR" program, which makes possible the reliability calculation in terms of market factors. Moreover, the model can be applied as an independent program for comparative evaluation of different ways for arrangement of the electricity and capacity markets in power systems considering different possible calculated states (modes) in the system.

Researches of the system states using the functional (3.19) and one of the methods for organizing the electricity market were carried out using the example of the same test circuit shown in Fig. 3.3 and the initial information specified in Tables 3.1, 3.2, 3.3 and 3.4. The calculations on effectiveness of different states (modes) of the interconnection reveal the possibility of quantitative validation of different control models and measures for reliability support under the market competition. The researches on the model allow the sets of rules for joint operation of private power companies to be laid down.

The wholesale price in the regional market arranged in accordance with the strategy of load coverage by the cheapest electricity was calculated to be equal to 0.0396 m.u./kW h, which coincided with the price offered by the first node.

The additional initial information is presented in Table 3.8.

The designed program was applied to research different strategies of assigning the rates and their influence on various types of power system effectiveness in different states (modes), both deficit and deficit-free. Some results of these studies are presented below.

Tables 3.9 and 3.10 demonstrate the calculation results of the normal states (modes) of the system that is represented by the test scheme. The performed calculation was compared with the calculations of other states (modes) and principles

Table 3.8 Additional cost parameters

Nodes of calculation scheme	1	2	3	4	5	6	7
Electricity price offered for wholesale market c_m^{el}, m.u./kW h	0.0396	0.0693	0.0528	0.0330	0.0528	0.0418	0.0528

of power system operation in the interconnection. For instance, the research was conducted on the expediency of constructing an own electricity source at node 7. It was found that, for the specific conditions under consideration, the construction of a power plant at this node is neither economically nor financially profitable if the cost of producing electricity at this station is higher than its price in the wholesale electricity market. Such a case is quite possible. In the example, moreover, there is an excess of capacity in the system. In this situation, it is more expedient to eliminate the deficit in node 7 by strengthening the connection of the node with the rest of the system. Quantitatively, it looks like this. Construction of the 50 MW power plant (with the corresponding decrease in the transfer capability of tie VII by 50 MW) causes decrease in the cost-effectiveness from 5124.76 to 5123.23 thousand m.u. in the system and from 149.96 to 147.93 at node 7. The financial effectiveness also falls in the system (from 125.74 to 125.09 m.u.), but rises at node 7 (from 1.26 to 1.90 thousand m.u.). (The indicated values are attributed to one hour of power system operation).

These complex relationships are also caused by variations in the wholesale price as a function of the formed electricity demand and supply (from 0.039 to 0.033 m.u./kW h in this case), which are described in the model. Assignment of higher values of f enhances cost-effectiveness and reduces capacity deficit at the corresponding node. The latter can be used to realize such strategies of limiting the consumers as their disconnection at the node with generation fall or, on the contrary, prohibition of electricity undersupply to consumers at this node. In the first case, f is decreased; in the second case, f is increased. In such a case, the coefficient f plays a role of the corresponding penalty.

In deficit state, the cost-effectiveness decreases and the profit often becomes negative both at the nodes with power deficit and without it.

The values of different types of efficiency (economic, financial) strongly depend on the excess (unplugged) capacity in EPS, as the formation of the price on the wholesale market depends on them. The more excess, the lower the price.

Thus, the considered model can be applied to a wide scope of researches on the effectiveness of different states (modes) under variation of the economic indicators and technical parameters.

Table 3.9 Calculation results of a normal calculate state (mode) of the interconnection (maximum load)

Characteristic	Power system							
	1	2	3	4	5	6	7	IPS
1. Generating capacity used to cover loads, MW	2021	1634	382	1349	508	700	0	6593
2. Load provided with coating, MW	2380	1500	500	100	1540	400	150	6570
3. Capacity deficit, MW	0	0	0	0	0	0	50	50
4. Capacity losses in ties, MW	10.8	0.87	1.83	1.03	11.63	0.72	0.20	26.27
5. Cost of electricity supplied to consumer, thousand m.u.	85.68	99.80	24.0	52.45	73.94	27.09	7.20	370.16
6. Damage caused by electricity undersupply, thousand m.u.	0	0	0	0	0	0	50	50
7. Costs on electricity production by node, thousand m.u.	75.50	56.34	15.30	13.50	15.25	24.32	0	200.21
8. Cost-effectiveness, thousand m.u.	1709.0	1068.6	359.7	36.5	1524.7	275.6	149.9	5124.8
9. Financial effectiveness, thousand m.u.	6.79	50.79	4.05	38.96	17.81	6.08	1.26	125.74

3.2.11 *Calculation of Energy Reliability Characteristics of Ties*

As is noted in Sect. 2.5, the energy reliability characteristics (ERChs) can be applied to reliability calculation of the schemes with practically any amount of nodes and ties via their step-by-step analysis.

The algorithm for ERCh calculation and application, which is used in the suggested model for reliability assessment of large power systems (see Appendix A.10), contains three fundamental errors degrading the calculation accuracy:

(1) At each successive step of calculations, the used ERChs are averaged for all hours of the year.;
(2) The accepted scheme of ERCh application does not take into account the correlation of capacity flows in diverse ties in a complex scheme.

Table 3.10 Components for assessment of the cost-effectiveness and financial effectiveness

Components, thousand m.u.	EPS							IPS
	1	2	3	4	5	6	7	
Cost-effectiveness including:	1709.0	1068.6	359.7	36.5	1524.7	275.6	149.9	5124.8
1. Prevented damage	1785	1125	375	50	1540	300	150	5325
2. Fixed costs	45.19	40	9.57	12.15	7.63	13.81	0	128.36
3. Fuel costs	30.31	16.34	5.72	1.35	7.62	10.50	0	71.85
Financial effectiveness, including:	6.79	50.79	4.05	38.96	17.81	6.08	1.26	125.74
1. Cost of electricity supplied to own consumers by own power plants	72.75	94.50	18.32	3.00	24.36	15.20	0	228.13
2. Cost of electricity supplied to own consumers from wholesale market	12.93	0	5.68	0	49.56	0	7.20	75.37
3. Cost of electricity supplied to wholesale market	0	5.30	0	49.45	0	11.89	0	66.64

(3) In the calculation of ERChs of ties between sufficiently large parts of the power system, the possible limitations on power flows in transmission lines in these parts of the system are not taken into consideration, as far as each part is represented in such a case as a "concentrated node."

However, the analysis of actual reliability calculations for different systems [27] showed that the total error in assessment of the reliability indices using ERC did not exceed 10%. Their application considerably decreased the time of reliability research for complex power systems and coordinated the reliability calculations for different territorial levels of the system.

3.2.12 Calculation of Reliability Indicators for the Functioning of Nodes and the System as a Whole

The reliability indices are calculated in the program on the basis of the values of the probabilities $q_{\text{def}\,m\varphi\beta k}$, $q_{\text{def syst}\,\varphi\beta k}$, and the capacity deficit $P_{\text{def}\,m\varphi\beta k}$, $P_{\text{def syst}\,\varphi\beta k}$ which are obtained in the previous blocks after calculation of each deficit state.

These data can be applied to determine the following reliability indices of power supply to consumers at nodes and in the system:

- the factors of failure-free power supply at each node and in the system at each interval (availability factors)—$\mathscr{P}_{ms}$, $\mathscr{P}_{\text{syst}\,s}$;
- the factors of failure-free power supply at each node and in the system during the year (availability factors)—$\mathscr{P}_{m}$, $\mathscr{P}_{\text{syst}}$;
- the average values of power undersupply at each node and in the system at each interval—$W_{\text{und}\,ms}$, $W_{\text{und syst}\,s}$;
- the average values of power undersupply at each node and in the system during the year—$W_{\text{und}\,m}$, $W_{\text{und syst}}$;
- the relative provision of power supply at each node and in the system at each interval—π_{ms}, $\pi_{\text{syst}\,s}$;
- the relative provision of power supply at each node and in the system during the year—π_{m}, π_{syst}.

Apart from the indicated reliability indices, which are the mathematical expectations (m.e.) of some random variables, the program enables to calculate the mean-square deviations of these random variables: $\sigma\,\mathscr{P}_{ms}$, $\sigma\,\mathscr{P}_{\text{syst}\,s}$, $\sigma\,\mathscr{P}_{m}$, $\sigma\,\mathscr{P}_{\text{syst}}$, $\sigma\,W_{\text{und}\,m\,s}$, $\sigma\,W_{\text{und syst}\,s}$, $\sigma\,W_{\text{und}\,m}$, $\sigma\,W_{\text{und syst}}$, $\sigma\,\pi_{ms}$, $\sigma\,\pi_{\text{syst}\,s}$, $\sigma\,\pi_{m}$, $\sigma\,\pi_{\text{syst}}$.

3.2.13 Processing of Calculation Results

The operation reliability indices for the calculated nodes and the system, as well as the energy reliability characteristics (ERChs), are determined as a result of calculations.

For convenient analysis of the results, the indicated indices are presented in a tabular form with the necessary wording.

3.3 Basic Characteristics of the "KORALL" and "YANTAR" Programs

The programs are intended to determine reliability indices of complex power system operation. Programs are designed, as was mentioned, in the algorithmic Fortran language. Brief description of the subprograms and the pattern of their interaction are presented in Appendix A (see Table A.3.1 and Fig. A.3.1).

The "KORALL" program was verified on the basis of the following calculations:

(1) calculations of the two- and three-nodal (of a "ring" type) test schemes manually and by the "KORALL" program;
(2) calculations of the test schemes by the "POTOK" [6] and "KORALL" programs;

(3) calculations of the eight-nodal scheme for comparison with the results obtained by the programs from Power Engineering Institute (PEI) [29] and Kyrgyz Research Department of Energy (Kyrgyz RDE) [30].

The results obtained manually and by the "KORALL" program coincided, and when comparing with the "POTOK" program, the results practically coincided (see Table 3.11).

The reliability indices for the eight-nodal scheme (for the initial data see Appendix C) were calculated by division of the initial scheme into some calculated subsystems with the subsequent use of ERCs to switch from one subsystem to another.

For the first calculation, the initial system was divided into two subsystems: subsystem I with nodes 1, 2, 5, 6, and 7 and subsystem II with nodes 3, 4, and 8 of the initial scheme. The obtained subsystems were taken as concentrated. As a result of the calculation, the ERChs of the tie between nodes 2 and 3 (2–3) of the original scheme were obtained (see Fig. C.1 of Appendix C).

The subsystem consisting of nodes 3, 4, and 8 of the initial scheme was calculated at the second stage including ERChs of the ties 2–3, and the new reliability indices of these nodes were obtained.

Then, the rest of the initial scheme was divided into two subsystems: subsystem III with nodes 1, 2, and 5 and subsystem IV with nodes 6 and 7 of the initial scheme. At first, the newly obtained subsystems were also assumed to be concentrated. And as a result of the calculation, ERC of ties 5–6 of the initial scheme was calculated. Afterward, the scheme consisting of nodes 1, 2, and 5 was calculated taking into account ERCs of ties 2–3 and 5–6. And finally, the scheme with nodes 6 and 7 was calculated at the last stage consisting of ERCh of the ties 5–6. Thus, the sought reliability indices were determined for all nodes of the initial scheme. The calculation results of one option of the scheme (Variant 2) are illustrated in Table 3.12.

Table 3.13 presents the results of reliability calculations for the considered scheme by the "KORALL" program in comparison with other programs.

Since the programs designed at SRPI and Kyrgyz RDE could calculate an expectation of the total undersupply only in the whole system, the results were compared using this index alone. The options of schemes 3 and 7 containing closed rings were calculated by the program designed at SEI (ISEM) only, as far as they could not be calculated by the programs of SRPI and Kyrgyz RDE.

Table 3.11 Results of calculations by the "POTOK" and "KORALL" programs

Reliability indices	Programs		Deviations
	POTOK	KORALL	
W_{und}	208.3	172.1	17.4
$\mathscr{P}$	0.94533	0.95889	1.4
π	0.99306	0.99446	0.14

Here: RI is reliability indices; K is program "KORALL"; P is program "POTOK"

Table 3.12 Results of reliability calculation for the eight-nodal test scheme of UPS. Variant 2

Number of node	$\mathscr{P}$	π	W_{und}	Rms deviation of π	Rms deviation of W_{und}
1	0.99650	0.99996	21.35	0.009892	530.71
2	0.98045	0.99472	70.4	0.022818	182.51
3	0.99454	0.99995	5.3918	0.009929	971.92
4	0.98486	0.99945	158.26	0.021316	4527.8
5	0.94968	0.99915	280.25	0.024229	8218.9
6	0.99994	0.99999	0.10437	0.000454	30.295
7	0.999999	0.999999	0.21×10^{-5}	0.000002	0.1362
8	0.99395	0.99991	5.2410	0.011129	677.56
UPS	0.94968	0.99966	441.0	–	–

Here: Rms deviation is reliability mean-square deviation

Table 3.13 Comparison of calculation results by the programs designed at SRPI, Kyrgyz RDE, and ISEM (SEI)

Variant	Transfer capabilities, MW								W_{und}, million kW h		
	Number of tie								SRPI	KygyzRDE	SEI
	1	2	3	4	5	6	7	8			
1	0	0	0	0	0	0	0	0	3048	3035	3044
2	2200	3000	1300	400	800	800	1000	–	438.5	445.3	441.0
3	2200	3000	1300	400	800	800	1000	1000	–	–	228.1
4	2200	3000	1300	4000	1800	800	1000	–	261.1	266.3	264.4
5	2200	3000	1300	400	1800	1800	2000	–	93.3	94.9	94.6
6	5000	5000	5000	5000	5000	5000	5000	–	35.1	33.7	34.3
7	5000	5000	5000	5000	5000	5000	5000	5000	–	–	33.75
8	4000	5000	2500	1000	2500	500	500	2500	–	–	33.75
9	System as one node (free flow area)								–	–	32.97

Table 3.13 shows that in Variant 7, the calculated value of the total undersupply is close to W_{und} for the system with the "unlimited" transfer capabilities and $q_{lin} = 0$ (see Variant 9). Variant 8 is an example of the UPS scheme adjustment with respect to transfer capabilities of the transmission network (in this case) based on the suggested approach and the information, which was applied to calculations of nodal reliability indices and ERChs of ties. Variant 8 maintaining the same reliability level as variant 7 offers an essential capacity saving (in MW km) in the network part of the system.

Different variants of the eight-nodal test scheme of UPS are described in detail in [31]. ERCp of one of the ties in the considered scheme is given as an illustration in Table 3.14. The example of this tie demonstrates capabilities offered by the knowledge of ERCs for engineering adjustment of the scheme.

Table 3.14 Energy reliability characteristic (ERCh) of tie 6 (South–North Caucasus) for variant 6 of the scheme

P, MW	$\mathcal{P}$	*P*, MW	$\mathcal{P}$	*P*, MW	$\mathcal{P}$	*P*, MW	$\mathcal{P}$
−5000	0	−2500	0.0125150	0	0.7855970	+2500	0
−4500	0	−2000	0.0117401	+500	0.0251558	+3000	0
−4000	0	−1500	0.0217570	+1000	0	+3500	0
−3500	0.0001930	−1000	0.0308160	+1500	0	+4000	0
−3000	0.0037650	−500	0.1027960	+2000	0	+4500	0

3.4 Application Spheres of the Programs

As was noted, the "KORALL" and "YANTAR" programs were intended mainly to research reliability of the main structure of power systems at the stages of designing and development planning.

Naturally, the key function of the programs is to analyze reliability of individual options of power system development or operation, which are worked out by some or other method. The indicated analysis can be used for solving such main problems as choice of the most reliable option from a set of them by comparative technical and economic evaluation and identification of "bottlenecks" in any researched power system scheme in the context of reliability.

In the current practice of design, the technical and economic analysis of different options is the central method for choosing the optimal path of power system development [32]. Reliability assessment of the researched variants, as was underlined, is an essential component of this analysis.

Assessment of the reliability indices for different options of power system development in combination with assessment of the discounted costs enables to choose the most promising option in the context of higher certainty. If the considered options are formulated so that the higher discounted costs without regard to the national damage due to power undersupply to consumers correspond to lower reliability, then the conditions for option choice are evident. If the discounted costs for different options are approximately equal, the power system development variant is chosen on the basis of the best value of the corresponding reliability index. Its best value is also used to determine power system development variant in the case of limited funds.

If the standards values of reliability indices are available, and if they are determined highly accurately, the variants must be adjusted so that the reliability indices characterizing them are equal. Then, the best variant is determined by the criterion of the least discounted costs (neglecting the damages caused by unreliability).

Finally, for integrated optimization of the main structure of power systems using the corresponding optimization program the designed subprogram can serve as a subblock of the refined reliability assessment of the structure under formation within this program. This subroutine can be used at each iteration of the optimization process, if computers have high speed. And also this subprogram can be used periodically in the development of the next variant of the development of the power system.

The comparative analysis of power system reliability will make possible both the choice of the best system development variant and the solution to the second mentioned problem. The latter is possible owing to the calculation of reliability indices for all nodes using the subprogram for the whole considered period and for its individual intervals. As a result, the operation conditions of energy areas and the pattern of their reliability change over time are evaluated. Such analysis reveals nodes with an insufficient level of power supply. Then, the extent of reliability improvement of both the indicated nodes and the resulting reliability of the whole system is estimated by different measures, since the system reliability depends on reliability of the nodes comprising it.

In power system reliability researches, the programs are supposed to be useful in two aspects:

(a) The researches of the effect of different factors of functioning, which determine power system reliability [33]. These studies will allow the quantitative reevaluation and readjustment of the known (or unknown) qualitative relationships among different conditions of system operation at different levels of power supply reliability achieved. Hence, the new methods and models for power system reliability assessment will be developed on a more reasonable basis.
(b) Investigation of the impact on the reliability of EPS of various design methods for the formation of the basic structures of EPS, conditions, and factors for planning the development of EPS (uncertainty of information, various ways of selecting the size and placement of reserves of generating capacity, repair planning, etc.).

In the indicated scope of problems with the expedient application of the "KORALL" and "YANTAR" programs, they, as the computational tools, should be compatible with different requirements determined by the specific character of the problem solved.

Thus, the use of the program "KORALL" or "YANTAR" for the needs of planning the development and design of EPS requires that many variants be calculated during the limited time. This fact makes the increased requirements for the time of calculation of one variant, but does not require an increased accuracy of the results obtained. In some cases, the low requirements for accuracy are explained by the application of insufficiently certain data or data obtained as a result of deep equivalenting of the initial information. On the contrary, in the research calculations, the requirements for the program speed are somewhat lowered, but the requirements for more complete and accurate calculations performed increase.

The developed programs allow to provide one or other of the indicated modes of their use through the appropriate organization of calculations and the assignment of the initial data (see Sect. 2.5).

The "YANTAR" program can be treated as a computational tool realizing a sufficiently full calculation model for the chosen reliability indices of power systems.

The calculation results can be applied to the engineering analysis and optimization of reliability of the power system and its energy areas.

The program can also be used as a reference one for testing and adjustment of the applied and newly constructed indirect (or simplified) models for reliability accounting in the control problems to be solved.

This program can be applied to solve the following problems:

- choice of all types of generating capacity reserves in the power system: reserves for scheduled repairs (current, medium, major), operating reserve, commercial reserve;
- rational allocation of the chosen size of the total reserve among energy areas and power plants of the power system taking into account the transfer capabilities of networks;
- calculation of power supply reliability for the energy areas (estimation of the probability of deficit-free operation, relative energy index of reliability, average volume of electricity undersupply to consumers over a year);
- assessment of power supply reliability for a concrete consumer in the considered power system;
- assessment of the system effect taking into account the reliability factor for different development variants of the main structure of power systems (generating and network components of the system and component of system is provision with primary energy resources);
- optimization of new equipment commissioning time in power systems using the reliability factor.

In addition to the key factors, it is possible to take into account such factors in the development and functioning of EPS as:

- the processes of equipment running-in and aging (via the corresponding assignment of statistical values of failure rate the equipment for the specified considered periods);
- special schemes for the operation of equipment (double blocks, special circuits for linking power lines, and the effect of internal ties of EPS on the intersystem ties);
- seasonal and other changes in the equipment set and characteristics of system elements including reliability characteristics;
- accounting for the time shifts in time for various nodes of EPSs and so on.

Evaluation of reliability indicators for various options for the operation of EPSs, combined with an estimate of the cost of the calculations, allows you to choose the best option in conditions of greater certainty.

For the recent decade, the "YANTAR" program has been applied to a large number of reliability calculations of diverse schemes: UPS of the USSR, UPS of Russia, IPS of East, the interstate interconnections of power systems of the Russian Far East–Japan; IPSs of Siberia, East and North China; IPSs of the Russian Far East, North Korea, South Korea, etc. The reliability of UPS of Russia was studied for construction of the 1150 kV transmission line on the territory of Russia from IPS of Siberia to Ural. Survivability of UES of Russia was assessed for large-scale and long-term contingencies (gas undersupply, NPP moratorium, isolated power system operation, low-water years of the Russian rivers, etc.). The calculations performed made it possible to take into account the reliability factor in the investigating of power system development and operation schemes [34, 35].

3.5 About Methodical Regulations on Reliability Accounting in Power System Development Planning

3.5.1 Justification of the Necessity of the Regulations and Their Purpose

Involvement of the integrated reliability assessment of the main structure of power system development variants in the engineering design calls for adjustment of some basic regulations of designing, clearer coordination of the problems with each other and with the complex problem. It is important that coordination of the problems in terms of the final integrated assessment of the variants will simplify partially the algorithms for solving some problems. However, at the same time, there arises a difficulty of the extensive practical application of the adjusted integrated reliability assessment of power system development variants.

The above said can organically be provided by the generation of the methodical regulations for reliability analysis in power system designing. Expediency of generating the indicated regulations follows from the widely recognized significance of the problem of maintaining reliable operation of future power systems.

On the other hand, the analysis of the current practice of reliability accounting in power system development planning reveals that despite the achieved progress in creation of the appropriate methods, algorithms, and programs, their practical application to designing falls behind. In various research and design institutions, this issue is often treated differently.

According to the authors, this gap is due to the following main reasons:

- not taking into account the nature and interrelationship of design tasks, as well as the features of design systems are not taken into account in the developed techniques. Sometimes, it makes to the practical impossibility of using the created algorithms and programs;
- unpreparedness of practitioners to apply the up-to-date methods for reliability assessment of complex systems;
- insufficiency and inefficiency of the efforts on implementation of the developed approaches.

The existing contradiction between the degree of the problem solution and practical application of the obtained results seems to be eliminated by issue of the methodical regulations (recommendations) on reliability accounting at power system development planning for the design institutions.

The successful control is known to depend substantially on the possibility and the degree of formalization and regulation of control actions. In this context, the methodical regulations and other similar guidance materials have always been an appropriate means for control quality improvement and unification, more distinct coordination of actions at different temporal and territorial levels of the hierarchical control.

This is testified, for example, by the extensive experience gained in power systems on preparation and issue of standard instructions, rules, methodological guidelines, and standards on different matters of operation and development, whose practical application had and has an essential organizational and technical-economic effect.

As concerns the considered problem, the methodical regulations can become a convenient practical tool for implementation of the developed and tested scientifically grounded methods, standards, and the corresponding automated computer facilities to solve the problems of system development optimization taking into account the reliability.

Coordination and mutual agreement of the optimization approaches and algorithms for different hierarchical levels, which are supported by the methodical regulations, will make it possible to adjust, partially simplify, and, above all, unify these approaches and algorithms even at the stage of their development.

The regulations will also help in the faster and extensive application of the techniques for reliability accounting by the personnel of design institutions, which generalize the results of the latest research, in formulation of different design documents (feasibility study, system development schemes, working projects, etc.).

The methodical regulations will contribute to a simpler review and adjustment of the calculation methods.

From the above said, it seems necessary to state the following scientific and organizational task: to prepare and publish the recommendations (preliminary) and then the methodical guidance on reliability accounting at power system expansion planning.

The work on recommendations is intended for the development of the regulations, whose application will ensure more complete and precise reliability accounting in power system planning.

However, the serious complications arising in the development of the regulations can be eliminated by the necessary researches to be conducted.

First of all, the generation of the recommendations and then the regulations calls for great efforts for their coordination with the design practice or for justification of changes in this practice, some principles of effective guiding materials, as well as for coordination of the regulations for the considered specific, territorial, time, and functional level of the control hierarchy with the similar regulations of other levels.

The methodical regulations should comprise algorithms (programs) for reliability accounting and guidelines for preparation and representation form of the input and output information.

Besides, the system of reliability standards, which is substantiated technically and economically, on the basis of the experience gained or in an expert way can be the real and, apparently, unique keystone for reliability accounting in some problems of power system development planning. In addition, the reliability standards should be coordinated by the territorial levels, time stages and power system generation and network parts.

3.5.2 Guidelines on the Content and Structure of Methodical Regulations

Specific features and capabilities of reliability accounting determine two main forms of the methodical regulations.

Whereas the system properties and sizes allow the formation of the unique approach to complex optimization of its development on the basis of all criteria including the reliability criterion, then the purpose of the methodical regulations is to fix ways and means for reliability accounting in the optimization models, as well as the nature of the harmonization of various norms and rules for different levels and stages of design.

When construction of unifying optimization models with sufficiently precise reliability accounting for the system proves to be impossible, the methodical regulations are intended to combine all the approaches for more complete consideration of the reliability factor, and advise an integrated, sufficiently accurate estimation of reliability levels of the future system development variants, which were obtained as a result of solving the design problems. Such an estimation model makes possible subsequent comparison of the variants to choose the best one or several based on a set of criteria including the reliability criterion (see Sect. 3.4).

When developing the methodical regulations, it is important to correctly understand that simultaneous optimization of the all links (components, parts, etc.) is not always possible for such large systems as power systems because of the problem dimension. Therefore, at the first stage of their development, it is worthwhile to split the global problem into individual solved groups of the optimization problems for the indicated the links.

The development of the structure of the methodical regulations is an important factor in their preparation.

From the formal standpoint and from the standpoint of the substance of the case, the methodical regulations can be developed only jointly by the research institutions and specialized design organizations.

3.5.3 Practical Experience in Development of the Methodical Regulations

In accordance with the presented recommendations, the Energy Systems Institute of the Siberian Branch of the Russian Academy of Sciences carried out the work on the development of the methodical regulations on reliability accounting in power system designing.

The principal emphasis was made on the main structure of power systems, as far as it is its most important element. The structure of the corresponding regulations was developed, and their first version was prepared at the Energy Systems Institute (SEI) [36]. The indicated structure, which was approved later by some competent organizations, is described in Appendix B.

Subsequently, the Krzhizhanovsky Power Engineering Institute and the organizations of "Energosetprojekt" were involved in the work on this problem. The following versions were developed jointly with them. Publication of "The Recommendations on reliability consideration in power system development planning" in 1979 was the result of the first stage of the work on the guidelines, which were approved by the Committee of the Scientific Council of the USSR Academy of Sciences for complex energy problems [37]. The prepared recommendations are a sufficiently useful material for further work in this area.

In order to work out the guidelines, it was necessary first of all to determine the sphere for the application of different approaches to power system development optimization, analyze the required level of modeling, determine the necessity of accounting some or other factors, and estimate the accuracy of the solutions obtained.

Incompleteness of separate works made it impossible to publish at the first stage a full version of the methodical regulations with inclusion of all necessary approaches to reliability consideration in designing, particular instructions on the appropriate spheres of taking different measures on reliability improvement and methods for technical and economic assessment of their appropriate application.

Of course, the publication of the methodical regulations will not imply an irrelevance to develop new approaches and algorithms. However, the corresponding technical requirements for their development should depend on the content of problems solved taking into account the reliability and their coordination within the developed regulations.

References

1. Rudenko YN, Cheltsov MB (1974) Reliability and redundancy in electro-energy systems. Methods of research. Science, Siberian Branch, Novosibirsk, 264 p (in Russia)
2. Lyalik GN, Urvantsev NP (1972) Determination of the emergency power reserve based on universal characteristics of the specific reserve, № 1, pp 26–30 (in Russia)

3. Nepomnyashchy VA (1975) Mathematical model of the reliability of the electric power system. In: Methodological issues of reliability research of large energy systems, issue 8. Reliability in the design and operation of energy systems. SEI SB AS USSR, Irkutsk, pp 110–118 (in Russia)
4. Nepomnyashchiy VA (1978) A calculation of reliability in the design of power systems. Energia, Moscow, 200 p (in Russia)
5. Itkin EA, Shadrin VA (1975) Model of reliability evaluation of the planned electric power system. In: Methodological issues of investigation of reliability of large energy systems, issue 8. Reliability in the design and operation of energy systems. SEI SB AS USSR, Irkutsk, pp 103–109 (in Russia)
6. Mogirev VV (1975) Algorithms and programs for calculating reliability indicators of electric power systems by the method of statistical modeling (program "Potok"). In: Methodological issues of investigation of reliability of large energy systems, issue 4. Mathematical methods of analysis of reliability. SEI SB AS USSR, Irkutsk, pp 24–35 (in Russia)
7. Itkin EA, Kolosov IS (1973) Estimation of the reliability of complex integrated power systems at the stage of their design. In: Methods for calculating the emergency reserve in complex power systems, issue 4. Proceedings of the Kirghiz research department of energy. Publishing house "Kyrgyzstan", Frunze, pp 21–27 (in Russia)
8. Vasin VP, Prokhorenko VI (1980) On the construction of the probability distribution function of the total power of the power system. Izv AN SSSR, Energy Transp 6:61–67 (in Russia)
9. Troshina GM (1978) On an approach to the solution of the problem of minimizing the deficiency of power in electric power systems. In: Methodological issues of investigating the reliability of large power systems, issue 15. SEI SB AS USSR, Irkutsk, pp 34–43 (in Russia)
10. Kovalev GF, Olenkevich VP, Patseva TV, Rudenko YN (1981) Algorithm and program for calculating reliability indicators of complex electric power systems by analytical method. In: Methodological issues of reliability research for large energy systems, issue 22. SEI SB AS USSR, Irkutsk, pp 17–32 (in Russia)
11. Kovalev GF, Olenkevich VP, Patseva TV (1981) Influence of factors of EES function and methods of their representation in models on the calculated value of the reserve of generating capacity. In: Methodological issues of reliability research of large energy systems, issue 23. SEI SB AS USSR, Irkutsk, pp 54–65 (in Russia)
12. Mogirev VV, Rudenko YN (1973) Algorithms for investigating the reliability of electric power systems of arbitrary configuration. Izv AN SSSR, Energy Transp 1:38–45 (in Russia)
13. Dikin II (1978) The method of interior points in mathematical programming. In: Applied mathematics. Science, Novosibirsk, pp 133–158 (in Russia)
14. Dikin II, Zorkal'tsev VI (1980) Iterative solution of problems for mathematical programming (algorithms for the method of interior points). Science, Novosibirsk, 148 p (in Russia)
15. Kovalev GF, Lebedeva LM (2000) A complex of models for optimizing the modes of calculated states in assessing the reliability of electric power systems. Preprint. ISEM SB RAS, Irkutsk, 73 p (in Russia)
16. Zorkaltsev VI, Kovalev GF, Lebedeva LM (2000) Models for estimating the power deficit of electric power systems. Preprint. ISEM SB RAS, Irkutsk, 25 p (in Russia)
17. Dikin I (1969) Convergence of an iterative process. In: All-union conference on problems of theoretical cybernetics (theses of reports). Science, Novosibirsk, pp 56–57 (in Russia)
18. Dikin II (1978) Application of algorithms for the method of internal points to minimize the power deficit in the electric power system. In collection: Methodological issues of research on reliability of large energy systems. Academy of Sciences of the USSR, SEI, Irkutsk, pp 23–33 (in Russia)
19. Zorkaltsev VI (1984) Relatively internal point of optimal solutions. In: A series of preprints "Scientific Reports", issue 100. Komi Branch of the UrB RAS, Syktyvkar, 48 pp (in Russia)
20. Dikin II (1976) Iterative solution of problems of linear and quadratic programming. Dokl Akad Nauk SSSR 174(4):747–748 (in Russia)
21. Yudin YB, Golshtein EG (1966) Linear programming. Mir, Moscow, 276 p (in Russia)
22. Fadeev DK, Fadeeva VN (1963) Computational methods of linear algebra. Fizmatgiz, Moscow, 468 p (in Russia)

23. Voropai NI, Kovalev GF, Lebedeva LM (1996) Efficiency of corresponding electric power systems in the interconnection during emergency and deficient conditions. In: PSCM, 4th international conference, 16–18 Apr 1996. Conference publication number 421. IEE, Savoy Place, London WC2, UK, pp 205–209 (in Russia)
24. Kovalev GF, Lebedeva LM (1996) Estimating the size of the reserves of energy and EPS power in market conditions. In: Social, technogenic and natural risk factors in production activities. Collection of reports. SEI of the SB RAS, Irkutsk, pp 86–96 (in Russia)
25. Kovalev GF, Lebedeva LM (1997) A model for calculating the efficiency of the regimes of merging electric power companies in a wholesale market. In: Methodological issues of reliability research for large energy systems, issue 49. Reliability of energy systems: economic and information aspects. St. Petersburg, pp 42–49 (in Russia)
26. Voropai NI, Kovalev GF, Lebedeva LM, Trufanov VV (1998) Modeling electric power system expansion under new conditions in Russia, vol 1. In: POWERCON '98, Beijing, 18–21 Aug, pp 720–724
27. Makarov AA (1998) Perfection of market relations in the electric power industry of Russia is an important condition for the development of the industry. Sb. Reports of the international congress of power engineers "Energy. Market: Integration", 25–26 June 1998, pp 149–154 (in Russia)
28. Kovalev GF, Mogirev VV, Syutkin NB et al (1984) About a complex of problems "Reliability" in ASDU of the united energy system of the USSR. In: Brief abstracts to the all-union scientific and technical conference "Questions stability and reliability of the power systems of the USSR", Tashkent, pp 62–63 (in Russia)
29. Volkov GA (1973) Determination of the reliability of power supply in integrated power systems using analytical methods of calculation. In: Proceedings of the III all-union scientific and technical meeting on the stability and reliability of the power systems of the USSR. "Energy", L., pp 518–527 (in Russia)
30. Itkin YA, Shadrin VA (1975) Construction of a model for analysis of the reliability of a complex electric power system using approximating curves for describing random variables. In: Methodological issues of reliability research for large energy systems, issue 12. SEI SB AS USSR, Irkutsk, pp 44–49 (in Russia)
31. Malkin PA, Kovalev GF (1981) Software for the calculation of reliability in the design of EPSs. In: Methods for investigating the reliability of large energy systems, issue 22. SEI SB AS USSR, Irkutsk, pp 7–16 (in Russia)
32. Rokotyan SS, Shapiro IM (eds) (1985) Handbook on the design of electric power systems. Energoatomizdat, Moscow, 352 p (in Russia)
33. Rudenko YN, Kovalev GF (1976) Methodical provisions for accounting for reliability in the design of electric power systems (project). SEI SB AS USSR, Irkutsk (in Russia)
34. Kovalev GF, Rudenko YN (1972) To the evaluation of the efficiency of the application of 1150 kV transmission lines in the main network of the EEES of the USSR at the level of 1980–90 from the point of view of reliability. Reports at the all-union conference "The main scientific and technical problems of electric power transmission with alternating current at a voltage of 1150 kV". "Nauka" Co., Novosibirsk, pp 110–121 (in Russia)
35. Manov NA (1976) Investigation of the accuracy of estimating the average annual indicators of reliability of a concentrated EPS for various forms of representation of electric power consumption modes in the calculation model. In: Methodological issues of investigating the reliability of large power systems, issue 7. SEI SB AS USSR, Irkutsk, pp 13–23 (in Russia)
36. Recommendations on accounting for reliability in the design of the development of electric power systems (1973) Academy of sciences of the USSR, Scientific council on complex energy problems, Moscow, 28 p (in Russia)
37. Belyaev LS, Kovalev GF, Podkavalnikov SV (1997) The efficiency of interstate electrical connections in the northern part of the Pacific region. Proc Russ Acad Sci Power Eng 6:3–10 (in Russia)

Chapter 4
Analysis of Impact of Some Factors of Functioning and Development of EPS and Methods of Representation of Them in Models on Calculated Indicators of Reliability

4.1 Objectives, Contents, and Main Provisions of the Research

The objectives of the investigation are:

(1) assessment of the impact of each condition (factor) on power system reliability level;
(2) analysis of the impact of representation forms and accuracy of these conditions in the calculation model on the sought indices;
(3) assessment of the impact of specific operation and development conditions of specific power systems on their reliability.

The first two groups of assessments are intended for verification of the developed methods and models. These researches make it possible to improve the effectiveness of the calculation models and simplify them reasonably.

The third group of the researches is intended for the practical analysis of effectiveness assessment of different engineering measures on reliability improvement of specific power systems (see Chap. 5). This Chapter will deal with the first two groups of the researches, which have a methodological character.

In the publications devoted to this problem (see Chap. 1), different authors express both coinciding conclusions and contradictory statements. Different recommendations in [1, 2] concerning the ways of load representation in the power system reliability models can serve as an example (see also Sect. 4.3.1). Therefore, these researches need continuation. It should be kept in mind that the impact of the factors is analyzed using different models, whose specific features can influence the results of the researches. For this reason, new researches of the factors and comparison of the results of these researches with the results of other researches remain topical. Finally, the quantitative estimates of the known (obvious) relations are of great practical importance to have clearer ideas about some properties and conditions of power system operation and development.

G. F. Kovalev and L. M. Lebedeva, *Reliability of Power Systems*, Power Systems, https://doi.org/10.1007/978-3-030-18736-1_4

As was noted in the previous Chapters, the power supply reliability depends mainly on the following factors of operation and development of individual subsystems comprising UPS:

- the volume and schedule of power consumption in the subsystems, which are represented rather completely by the load schedules and accuracy of its forecast;
- the level of equipment emergency rate, which depends on its composition and structure, the degree of its running-in or aging, the accepted policy of scheduled repairs and some other factors;
- the volume and schedules of reserve utilization in each subsystem. The actual reserve utilized in each subsystem consists of the internally available reserve and the power that can be transmitted from other subsystems as mutual assistance. The latter, in general, depends on the share of available capacity, which can be made available for the subsystem in the corresponding operating conditions, and on the transfer capabilities of transmission lines forming a power system network.

In order to assess the impact of the mentioned factors on reliability indices, it is important to determine the parameters, which would characterize these factors in a generalized way.

Such parameters can be the following:

(a) power consumption parameters:

- the maximum load $P^1_{\max}$;
- the power generation (required) W or, possibly, the peak load utilization hours $T^1_{\max}$;
- the error in the peak load forecast $\bar{\rho}$ (p.u. of $P^1_{\max}$);
- mean square deviation (m.s.d.) of the random load fluctuations σ^1;

(b) the relative value of the mathematical expectation (m.e.) of the capacity in an emergency state

$$\overline{\mathrm{M}}P^{\mathrm{g}}_{\mathrm{emerg}} = \bar{q}_{\mathrm{emerg}} = \sum_{j=1}^{J_{\mathrm{r}}} P^{\mathrm{g}}_j I_j q^{\mathrm{g}}_j / \sum_{j=1}^{J_{\mathrm{g}}} P^{\mathrm{g}}_j I_j, \tag{4.1}$$

which characterizes reliability of the capacity installed in the system;

(c) the reservation conditions:

- the calculated (for the period of peak load) value of the operating reserve P_{op}, or its relative value

$$r_{\mathrm{op}} = P_{\mathrm{op}}/P^1_{\max} = \left(P^{\mathrm{g}}_{\mathrm{instal}} - P^{\mathrm{g}}_{\mathrm{repair\,reserv}} - P^1_{\max}\right)/P^1_{\max}; \tag{4.2}$$

- ERChs of the ties, which comprehensively characterize reservation capabilities of the transmission lines adjacent to the subsystem. The characteristics should take into account their transfer capabilities, durations of their outages in scheduled and emergency repairs and available free capacities in the system at its different states (modes).

The following two Sections present the most essential results of estimating the quantitative dependences of the calculated values of reliability indices on the indicated factors of power system operation and development, on the representation forms and accuracy of these factors in the developed model.

4.2 Analysis of the Impact of Operating Conditions and Development of EPS on the Level of Reliability of Electricity Supply

The degree of materiality and influence is considered for each of the conditions.

The significance criterion of this or that factor is taken as a degree of change in the sought reliability indices calculated with the factor in comparison with the values of these indices calculated without this factor. The deviation by the value above 5–10% (an admissible error in engineering calculations [3]) is the criterion of the factor significance in the research.

The criterion of "strength" (degree) of the influence of the factor considered is the degree of relative change in the values of calculated reliability indicators in comparison with the corresponding relative changes in the quantitative changes of the factor itself.

The relative changes in the factors (FF) and the sought reliability indices (RI) are determined by the formulas:

$$\Delta\overline{\mathrm{FF}} = \frac{\mathrm{FF}_2 - \mathrm{FF}_1}{\mathrm{FF}_1} \times 100\% \quad \Delta\overline{\mathrm{RI}} = \frac{\mathrm{RI}_2 - \mathrm{RI}_1}{\mathrm{RI}_1} \times 100\%. \tag{4.3}$$

It is assumed that if the relative change of the considered reliability index is of the same or a higher order than the change of the factor, the impact of this factor on this index should be treated as sufficiently strong. The relative change of the reliability indices of a lower order than the change of the factor itself indicates an insignificant impact of this factor (condition).

The analysis of each of the factors was carried out taking into account all the others in the most complete and accurate form (with the exception of random load deviations, the neglect of which in the basic model obviously did not change the principal relations, but significantly reduced the calculation time).

The researches revealed that the response of different reliability indices to changes in the power system conditions of operation and development in terms of the chosen criteria is formally different. In this context, the indices $\mathscr{P}$ and π, in some cases become uninformative, since they change in the 3rd–5th decimal places with an objectively significant change in the level of reliability of the EPS studied. Therefore, the index W_{und}[1]—power undersupply—is considered in the work as the main index because it is best suited and descriptive for the corresponding comparisons on the basis of the chosen criteria. The indices $\mathscr{P}$ and π play, in this case, a secondary role.

The calculation results of some abstract systems consisting of similar units of identical units, which cover the fixed load, are applied in parallel with the calculation results of the UPS development options (see Appendices C and D) to reveal the corresponding dependences. Besides, the results of other power system studies including in some cases the researches of the concentrated systems, since in principle some dependences can keep their nature and quantitative characteristics despite the power system complexity. In such a case, the representation of regularities by the example of a complex power system becomes needless, since this representation becomes invisible.

The most typical calculations are chosen for illustration from a great number of calculations made for systems with different operation conditions. Several power systems substantially differing in power consumption level, set of units, reservation level, etc., are researched to improve the evidence and demonstrate the similarity of the obtained results and conclusions in the analysis of some or other factors.

The initial data of the researched power systems are given in Appendices 3 and 4.

4.2.1 Power Consumption Schedules

On the load amount and its schedules shape.

Table 4.1 presents the calculation results, which show that the power systems (IPS is taken as an example and the initial data see Appendix D) are characterized by the sufficiently high "natural" reliability (see the values of π in items 1 and 2 of Table 4.1), i.e., the reliability, when the measures on its improvement such as formation of operating reserve and assistance of other subsystems are not taken.

The use of the measures (see items 3–6 of Table 4.1) substantially raises the power supply reliability level.

Comparing the calculated reliability indices for the calculated conditions 1 and 2 in Table 4.1, it is particularly important to underline an essential system reliability improvement because of seasonal and daily valleys in power system load schedules. The relative value of power undersupply in this case is 0.3–0.6 of the undersupply for $P^1(t) = P^1_{max} = \text{const}$. Hence, this fact increases the need to take into account a real character of load as an essential factor (different calculation methods for load accounting are compared in the next Section).

Table 4.2 presents the power supply reliability levels as a function of power consumption levels in different systems (the initial data are given in Appendix C) under otherwise equal conditions. The presented values of reliability indices clearly show the need to take into consideration the error in load forecasting at power system

[1]Power undersupply is expressed in million kW·h throughout the work.

Table 4.1 Reliability levels of power system operation in different operation conditions by the example of IPS 1

No	Calculated conditions of power system operation	Indices	Variants of equipment emergency rate		
			Optimistic	Pessimistic	Medium
1.	Load schedule $P^1(t) = P_{\max^1} = \text{const.}$ Isolated system operation. Reserve $r_{op} = 0$	W_{und}	12,421	30,601	21,499
		$\Delta \overline{W}_{und}$	Higher by 62,200 times	Higher by 278 times	Higher by 2460 times
		π	0.96840	0.92156	0.94500
2.	Actual load schedule. Isolated system operation. Reserve $r_{op} = 0$	W_{und}	2460	12,012	6329
		$\Delta \overline{W}_{und}$	Higher by 12,300 times	Higher by 109 times	Higher by 695 times
		π	0.98980	0.95000	0.97370
3.	Same, but $r_{op} = 0, 03$	W_{und}	458	5483	2102
		$\Delta \overline{W}_{und}$	Higher by 2300 times	Higher by 49 times	Higher by 230 times
		π	0.99810	0.97720	0.99127
4.	Same, but $r_{op} = 0, 06$	W_{und}	43	1794	425
		$\Delta \overline{W}_{und}$	Higher by 215 times	Higher by 17 times	Higher by 46 times
		π	0.99982	0.99256	0.99824
5.	Same, but $r_{op} = 0, 08$	W_{und}	7	726	119
		$\Delta \overline{W}_{und}$	Higher by 34 times	Higher by 5.6 times	Higher by 12 times
		π	0.99997	0.99700	0.99950
6.	Actual load schedule. System operates in parallel with other IPSs. Reserve $r_{op} = 0, 06$	W_{und}	0.2	109.7	9.1
		$\Delta \overline{W}_{und}$	–	–	–
		π	0.99999	0.99954	0.99996

Variant 1, first level

development planning. For example, the deviation of $P^1_{\max}$ by 2–5% of its intended value changes the value of W_{und} by more than 2–10 times.

Calculations of multinode schemes showed that the effect of the error of the forecast of power consumption of separate energy nodes on their reliability, although it retains the nature noted above, but it somewhat is weakened due to the actions of such factors as: decrease in the relative magnitude of the forecast error as the absolute level of consumption in the EPS increases; the possibility of a multi-valued

Table 4.2 Calculation results of power system reliability indices with change in power consumption levels

$\Delta \overline{P}^{l}_{max}$, %	−10	−5	−2	0	2	5	10
IPS 5	$P^{l}_{max} = 43900$ MW, $r_{op} = 1,025$ %						
W_{und}	2.1148	142.4	800.82	1844.7	3458.3	6893.0	15073.0
ΔW_{und}, %	−99.885	−92.281	−56.588	–	+87.472	+273.665	+717.1
$\mathscr{P}$	0.99952	0.97564	0.89648	0.82576	0.74751	0.63726	0.46421
π	0.99999	0.99956	0.99759	0.99456	0.99001	0.98065	0.95961
IPS 1	$P^{l}_{max} = 21350$ MW, $r_{op} = 13,3$ %						
ΔW_{und}	0.00001	0.10266	0.93155	3.358	10.898	51.874	434.53
ΔW_{und}	–	−96.963	−72.26	–	+224.54	+1445	+12,840
$\mathscr{P}$	0.99999	0.99996	0.99964	0.99889	0.99686	0.98613	0.91650
π	0.99999	0.99999	0.99999	0.99998	0.99994	0.99970	0.99763
IPS 6	$P^{l}_{max} = 8750$ MW, $r_{op} = 6,68$%						
ΔW_{und}	0.11466	2.4975	12.353	31.626	75.55	228.26	883.44
ΔW_{und}	−99.637	−92.103	−60.94	–	+138.89	+621.75	+2693
$\mathscr{P}$	0.99990	0.99806	0.99206	0.98152	0.95904	0.89885	0.74648
π	0.99999	0.99996	0.99982	0.99952	0.99889	0.99673	0.98794

forecast error in different subsystems; mutual assistance between the nodes. Thus, the interconnection of power systems, in this case, is of a certain benefit.

The calculations of reliability indices for three nodes forming a ring scheme are presented in Table 4.3 as an example. In each node $P^{g}_{avail} = 16000$, $q^{g} = 0.05$; $P^{l}_{max} = 15000$, the load schedules correspond to the actual ones; the ties between the nodes consist of two transmission lines with a transfer capability of ± 1000 each, $q^{tl} = 0.01$. The error in power consumption forecast was assumed to be 1%, and the loads at nodes 1 and 2 proved to be 1% higher, and at node 3–1% lower.

Table 4.4 illustrates the dependence of power system reliability indices on the degree of power supply irregularity, which is represented here by the number of peak load utilization hours T^{l}_{max} with $P^{l}_{max} =$ const. The main dependence is obvious a priori: the lower is T^{l}_{max}, the higher is the calculated reliability of power supply to consumers under otherwise equal conditions. The calculations also show that this dependence is nonlinear and that the extent of power undersupply change is much higher than the extent of change of T^{l}_{max}. And, this relationship preserves its character at any levels of P^{l}_{max} and r_{op} (see Table 4.4).

However, it should be pointed out here that different shapes of the load schedules can correspond to the same T^{l}_{max} and P^{l}_{max}. Figure 4.1 presents three different load schedules with $T^{l}_{max} = 7750$ has an example. The values of reliability indices corresponding to the load schedules in Fig. 4.1 are given in Table 4.5.

Table 4.3 Example of calculations of power supply reliability indices with change in consumption levels in the three-nodal scheme

Indices	Nodes	Variants		
		Initial	With deviations from forecast	
		RI	RI	$\Delta\overline{\text{RI}}$, %
W_{und}	1	98.264	119.22	21.3
	2	98.264	119.22	21.3
	3	98.264	118.25	19.2
	System	294.79	355.39	20.6
$\mathcal{P}$	1	0.97865	0.95317	2.6
	2	0.97865	0.95317	2.6
	3	0.97865	0.95319	2.6
	System	0.97864	0.95316	2.6
π	1	0.99914	0.99896	0.02
	2	0.99914	0.99896	0.02
	3	0.99914	0.99896	0.02
	System	0.99914	0.99896	0.02

Table 4.4 Calculation results of dependences $\text{RI} = f\left(T^1_{\max}\right)$

$T^1_{\max}$, h	7000	7250	7500	7750	8000	8250	8500	8760
$\Delta\overline{T}^1_{\max}$, %	−9.69	−6.46	−3.23	0	+3.23	+6.46	+9.69	+13.05
IPS 5	$P^1_{\max} = 43900\,\text{MW},\ r_{\text{op}} = 1,025\%$							
W_{und}	1263.5	1389.8	1655.2	1844.7	2312.2	3221.9	5117.2	8212.8
ΔW_{und}, %	−31.5	−24.7	−10.3	–	+25.3	+72.5	+177.4	+345.2
$\mathcal{P}$	0.99920	0.99430	0.93560	0.89648	0.80935	0.64511	0.38315	0.19047
π	0.99997	0.99951	0.99902	0.99759	0.99392	0.98678	0.97197	0.94750
IPS 1	$P^1_{\max} = 21350\,\text{MW},\ r_{\text{op}} = 13,3\%$							
W_{und}	2.153	2.304	2.648	3.358	4.198	5.533	10.055	28.850
ΔW_{und}, %	−35.9	−31.4	−21.1	–	+25.0	+64.8	+199.4	+759
$\mathcal{P}$	0.99998	0.99931	0.99893	0.99889	0.99740	0.99573	0.99071	0.97692
π	0.99999	0.99993	0.99981	0.99998	0.99996	0.99992	0.99982	0.99937
IPS 7	$P^1_{\max} = 7500\,\text{MW},\ r_{\text{op}} = 31.4\%$							
W_{und}	0.29×10^{-4}	0.31×10^{-4}	0.36×10^{-4}	0.4×10^{-4}	0.49×10^{-4}	0.68×10^{-4}	1.26×10^{-4}	2.86×10^{-4}
ΔW_{und}, %	−39.9	−22.8	−11.3	–	+23	+68.8	+214	+615
$\mathcal{P}$	0.99999	0.99999	0.99999	0.99999	0.99999	0.99999	0.99999	0.99999
π	0.999999	0.999999	0.999999	0.999999	0.999999	0.999999	0.999999	0.999999

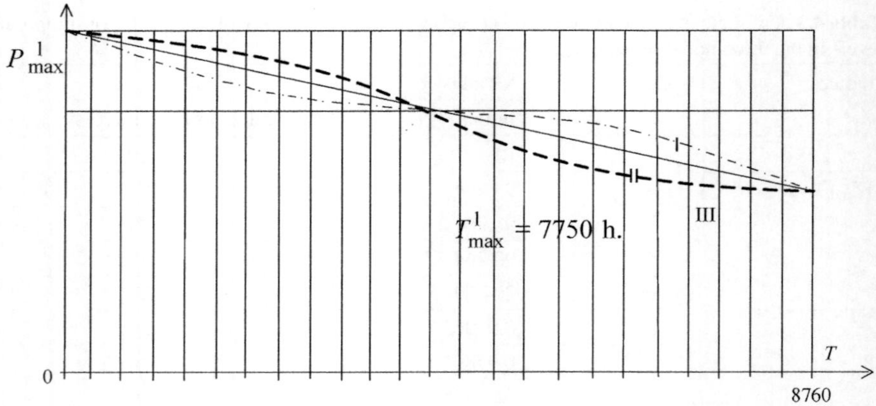

Fig. 4.1 Some possible configurations of power system load duration schedules

Table 4.5 Calculation results of reliability indices for three load schedules shapes presented in Fig. 4.1

Indices	Number of schedule		
	I	II	III
IPS 5			
W_{und}	1383.5	1844.7	2582.6
$\Delta \overline{W}_{und}$, %	−25	0	+40
$\mathscr{P}$	0.90052	0.82576	0.74798
π	0.99820	0.99456	0.98884
IPS 1			
W_{und}	2.384	3.358	5.608
$\Delta \overline{W}_{und}$, %	−29	0	+67
$\mathscr{P}$	0.99897	0.99889	0.99652
π	0.99999	0.99998	0.99988
IPS 8			
W_{und}	0.28×10^{-4}	0.4×10^{-4}	0.68×10^{-4}
$\Delta \overline{W}_{und}$, %	−30	0	+70
$\mathscr{P}$	0.99999	0.99999	0.99999
π	0.999999	0.999999	0.999999

It follows from them that the schedules' shape is a sufficiently important factor, which should be taken into account in the adjusted calculations of power system reliability by setting the load schedules by their actual shape.

The calculations show that the presented dependence does not change in a complex power system with weak ties.

On the random load fluctuations. The impact of random ("irregular") load fluctuations on power system reliability is confirmed by the calculations, and the results

Table 4.6 Calculation results of power system reliability indices with random fluctuations (A) and without them (B)

Type of calculation	A	B
IPS 5	$P^1_{max} = 43900$ MW, $\sigma^1 = 0, 022$, $r_{op} = 1, 025\%$	
W_{und}	2202,4	1844,7
$\Delta\overline{W}_{und}$, %	–	−16,24
$\mathscr{P}$	0,82241	0,82576
π	0,99351	0,99456
IPS 1	$P^1_{max} = 21350$ MW, $\sigma^1 = 0, 031$, $r_{op} = 13, 3\%$	
W_{und}	13,711	3,358
$\Delta\overline{W}_{und}$, %	–	−75,51
$\mathscr{P}$	0,999636	0,99889
π	0,99992	0,99998
IPS 7	$P^1_{max} = 8750$ MW, $\sigma^1 = 0.048$, $r_{op} = 6, 86\%$	
W_{und}	124,56	31,626
$\Delta\overline{W}_{und}$, %	–	−74.6
$\mathscr{P}$	0,95227	0,98152
π	0,99813	0,99952

of which are summarized in Table 4.6. In these calculations, the random fluctuations were presented by the normal distribution law, which is approximated by seven stages (probabilities: 0.006; 0.06; 0.242; 0.384; 0.242; 0.06; 0.006) with a range of $\pm 3\sigma^1$. The relative error in assessment of W_{und} is determined by formula (4.3, see above).

As is seen from Table 4.6, neglect of random load fluctuations in determination of the calculated reliability of real power systems results in the error reaching in individual cases 75%. Thus, in the general case, this factor should be considered in the analysis of power system reliability (Sect. 4.3 discusses the ways to do it better).

However, it should be pointed out here that the error considerably decreases at the lower values of σ^1.

It is known that σ^1 is usually the lower, the higher is P^1_{max} [3, 4]. On the whole, the accounting of all factors determining *power consumption* schedules enables to draw the following conclusions.

1. These factors are very significant, and therefore, to refine reliability calculations, one must strive for their fairly complete and accurate accounting, since the failure to account in general or the error (even not exceeding 5–10%) in setting the initial information on power consumption entail the achievement of desired reliability indicators, the error of which can significantly exceed the engineering allowable 5–10%.
2. On the other hand, the power system reliability calculations in real operation conditions show that in some cases it is possible to simplify calculations of

power supply reliability. For example, the impact of the error in forecasting and the random fluctuations of load on the assessment of power system reliability level decrease with the growing power consumption and the rising reservation level.

4.2.2 Composition, Structure, and Parameters of Generating Capacity

Composition and structure of generating capacity. The impact of the composition and structure of installed capacity on reliability indices was assessed based on the analysis of the results of reliability calculations for an idealized system. In this case, just an abstract system reveals most vividly the dependence interesting to us that is observed in real systems as well.

The analysis shows that somewhat interesting results are obtained in the research of the reliability indices as a function of the amount of units in the system under otherwise equal conditions, namely $P^{\mathrm{g}}_{\mathrm{avail}} = \mathrm{const}$; $P^{\mathrm{l}} = P^{\mathrm{l}}_{\max} = \mathrm{const}$.

For example, for the same available capacity formed by the different amounts of I homogeneous units $\left(I = \frac{P^{\mathrm{g}}_{\mathrm{aval}}}{P^{\mathrm{g}}_{\mathrm{agr}}} = \mathrm{var}\right)$, the same value of q and the same amount of operating reserve

$$r_{\mathrm{op}} = \frac{P^{\mathrm{g}}_{\mathrm{avail}} - P^{\mathrm{l}}_{\max}}{P^{\mathrm{l}}_{\max}} = \frac{I \cdot P^{\mathrm{g}}_{\mathrm{unit}} - B \cdot P^{\mathrm{g}}_{\mathrm{unit}}}{B \cdot P^{\mathrm{g}}_{\mathrm{unit}}} = \frac{I - B}{B} = \frac{R}{B}$$

the calculations made on the basis of conventional analytical expressions for reliability indices

$$\left.\begin{aligned} \mathscr{P} &= 1 - \sum_{i=R+1}^{I} C_I^i p^{I-i} q^i = \sum_{i=0}^{R} C_I^i p^{I-i} q^i; \\ W_{\mathrm{und}} &= T \cdot P^{\mathrm{g}}_{\mathrm{unit}} \sum_{i=R+1}^{I} C_I^i p^{I-i} q^i (i - R) \\ \pi &= 1 - \frac{1}{B} \sum_{i=R+1}^{I} C_I^i p^{I-i} q^i (i - R), \end{aligned}\right\} \tag{4.4}$$

which are obtained under these conditions reveal that with the increasing amount of I units the pattern of change in the reliability indices $\mathscr{P}$ and π is different for different levels of generating equipment emergency rate (see Fig. 4.2). In general, the system reliability level naturally increases with the increase of I, as evidenced by the index π.

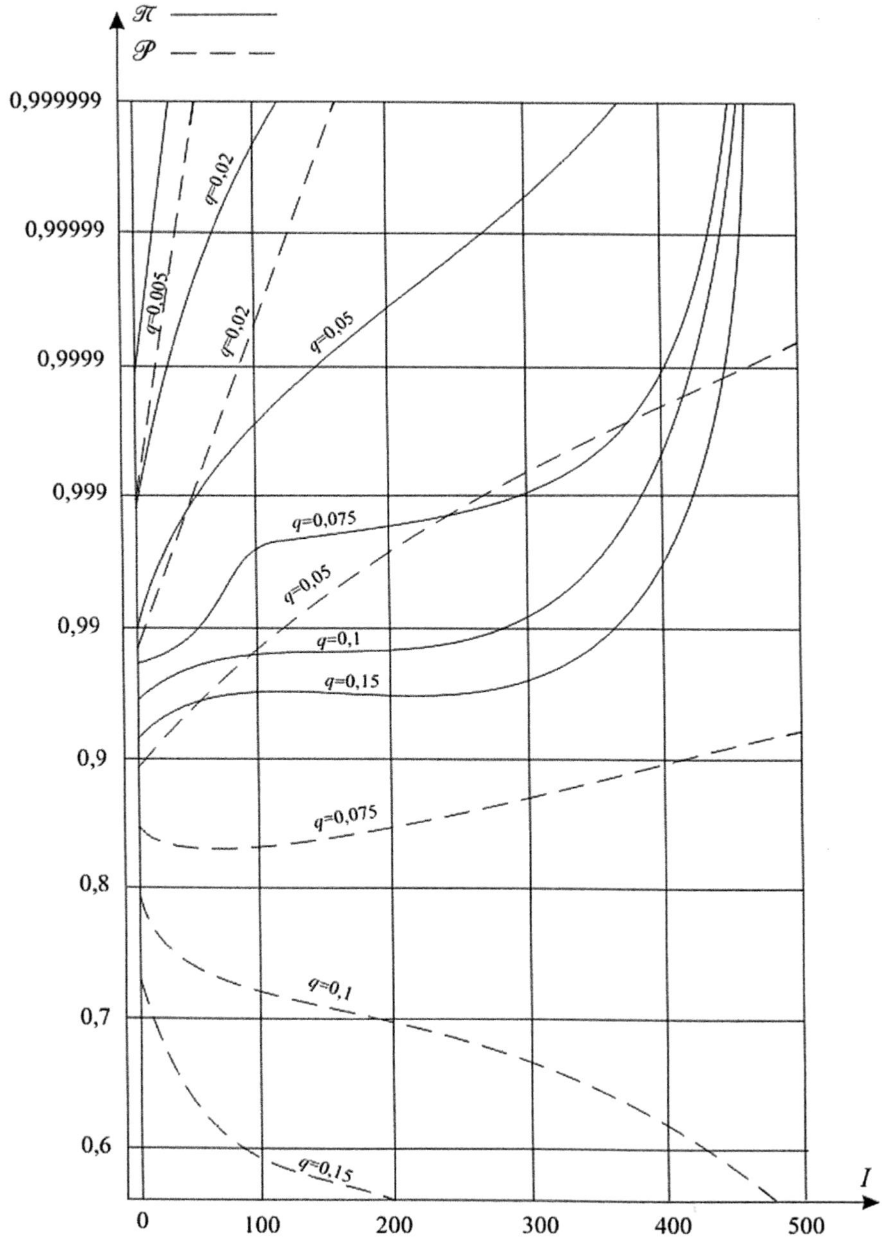

Fig. 4.2 Reliability indices as a function of the amount of units installed in the system $\left(P^{g}_{avail} = \text{const}, P^{1}_{max} = \text{const}, r_{op} = 0.1\right)$

However, at the same time, such operating conditions of EPS are formed that the $\mathscr{P}$ index in some cases sharply decreases with increasing number of units with high accident rate ($q > 0.08$) and with low values of capacity reserves $\left(r_{\text{op}} < 0.1\%\right)$.

The reason is that the increase of q and I leads to an increase of the relative duration of system operation with a large number of units under simultaneous emergency outage. This is seen from the schedules of the dependence of probability that all system units operate depending on I and q (Fig. 4.3). As a result, the duration of system operation with a certain (even though relatively low) shortage increases. Naturally, the values of the index $\mathscr{P}$, which do not depend on the shortage level, decrease.

The corresponding calculations show that the revealed dependences also hold for systems with the actual load schedules $\left(P^{1} \neq \text{const}\right)$ at the low values of r_{op} and the higher emergency rate q of equipment.

On the whole, the presented schedules make it possible to conclude that in terms of the reliability for the real values of q, r_{op} and I the power system should include a great amount of units with a lower capacity than vice versa. Thus, the quantitative characteristic of the contradiction (known to engineers) between the economic efficiency and reliability of power system operation with power units of higher capacity is confirmed. In addition, sometimes, the contradiction is aggravated by the higher emergency rate of these units.

Naturally, the constructed models for power system reliability assessment take account of the composition and structure of generating capacity as one of the most essential factors. It is important solely that any form of accounting these factors (see Sect. 4.3) would display the revealed dependences correctly enough.

Emergency rate of equipment. Figure 4.4 presents the $\mathscr{P}$ and π dependencies on the emergency rate q of equipment with the remaining unchanged parameters $\left(r_{\text{op}}, P^{1}, P^{\text{g}}_{\text{avail}}\right)$ for an idealized system [see formula (4.4)].

Similar relations hold for real systems as well (see Fig. 4.5). In this case, $\bar{q}_{\text{emerg}}$ is taken as a parameter of the emergency rate of equipment [see formula (4.1)]. A less smooth change in the reliability indices for IPS 1 in Fig. 4.5 is explained by the fact that the same value of $\bar{q}_{\text{emerg}}$ can be obtained at different changes in the emergency rate of the groups of units installed in the system which leads to the corresponding variations of power system operation reliability level.

Table 4.7 illustrates a quantitative dependence of the calculated reliability indices on the emergency rate $\bar{q}_{\text{emerg}}$ of equipment (4.1) exemplified by two highly different IPSs (the initial data are given in Appendix C).

The changes of W_{und} reveal its substantial dependence on reliable operation of generating equipment in the power system. In the generalized form, the value of $\bar{q}_{\text{emerg}}$ reflects the structure of generating capacities, dynamics of new equipment commissioning and forecasts of equipment operation level from the standpoint of its emergency rate.

Table in Appendix D.12 presents the calculated values of $\bar{q}_{\text{emerg}}$ for two cases: the hypothetical example, when the equipment emergency rate during its running-in does not differ from the emergency rate $\bar{q}_{\text{emerg no-run}}$ under normal operation and for

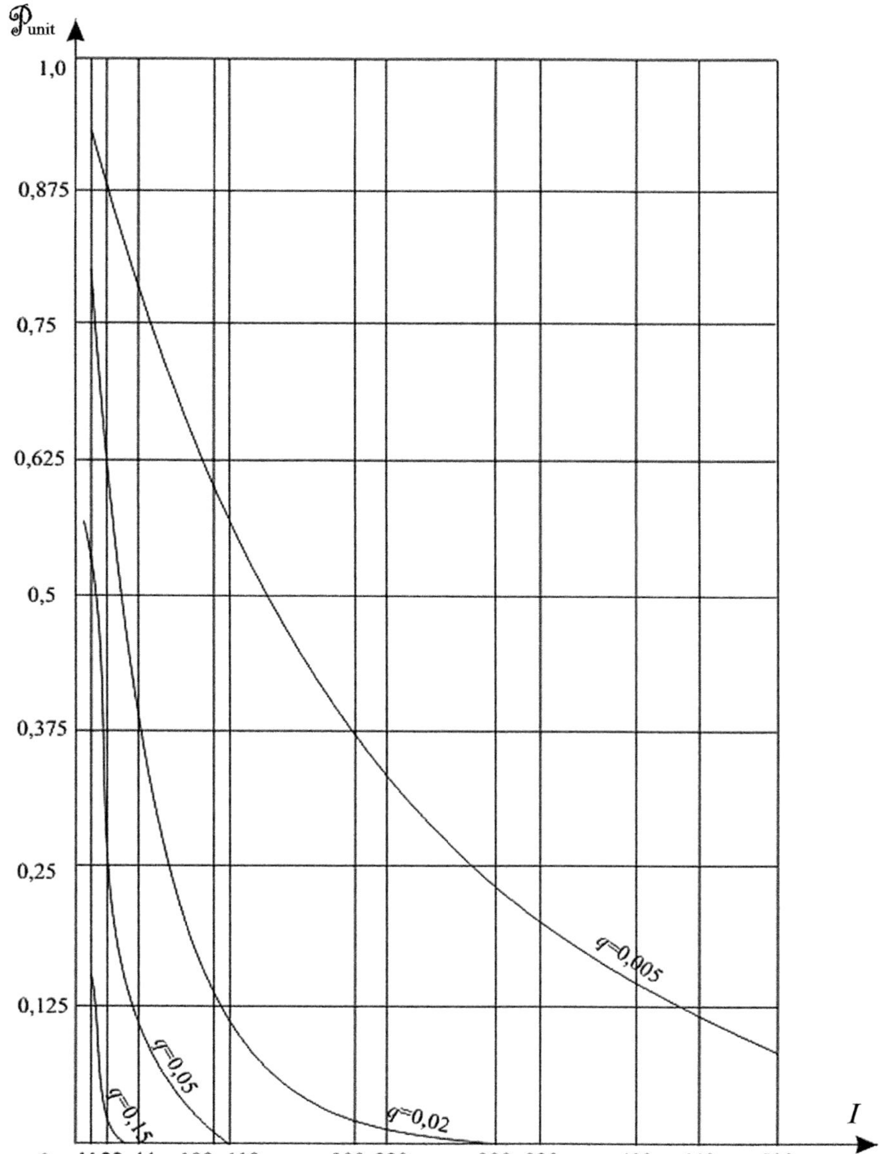

Fig. 4.3 Probability operation of all units $\mathscr{P}_{unit}$ as a function of their amount I and the emergency rate q

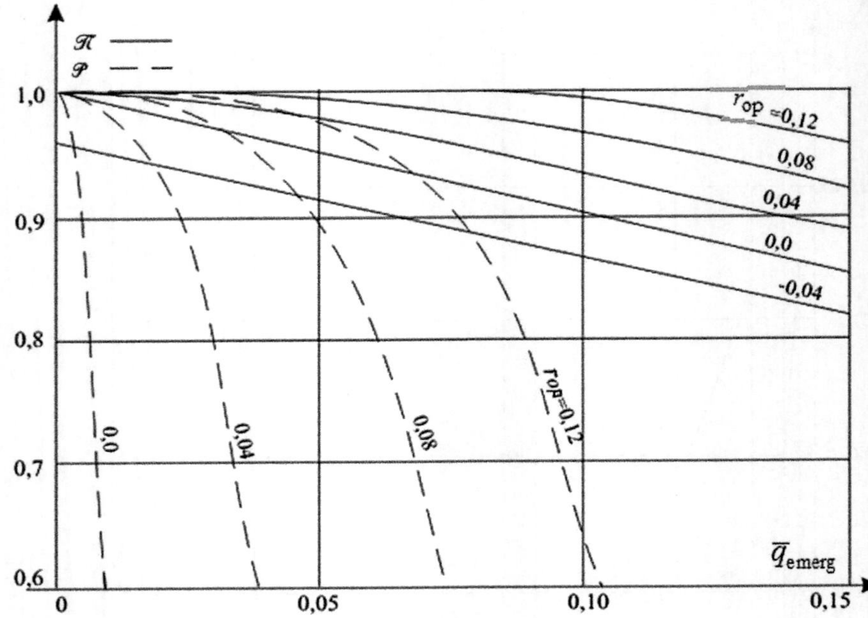

Fig. 4.4 $\pi = f_1(\bar{q}_{emerg})$ and $\mathscr{P} = f_2(\bar{q}_{emerg})$ for r_{op} = const

Table 4.7 Calculation results of power system reliability indices at different values of the generalized unavailability factor of generating equipment

$\Delta\bar{q}_{emerg}$, %	−50	−20	−10	0	10	20	50
IPS 3							
$\bar{q}_{emerg}$	0.0184	0.0294	0.0331	0.0368	0.0404	0.0441	0.0551
W_{und}	198.95	464.96	578.59	704.82	843.62	1388.9	2028.9
$\Delta\overline{W}_{und}$, %	−71.7	−34	−18	–	+20	+97	+187.5
$\mathscr{P}$	0.94764	0.90966	0.89663	0.88348	0.87027	0.84324	0.80269
π	0.99906	0.99781	0.99727	0.99667	0.99602	0.99345	0.99043
IPS 6							
$\bar{q}_{emerg}$	0.0184	0.0294	0.331	0.0368	0.0404	0.0441	0.0551
W_{und}	9.002	31.164	42.622	56.424	72.698	91.596	165.34
$\Delta\overline{W}_{und}$, %	−84	−44.7	−24.3	–	+29.1	+62.5	+193.5
$\mathscr{P}$	0.99263	0.98064	0.97540	0.96957	0.96316	0.95621	0.93230
π	0.99986	0.99953	0.99936	0.99915	0.99891	0.99862	0.99752

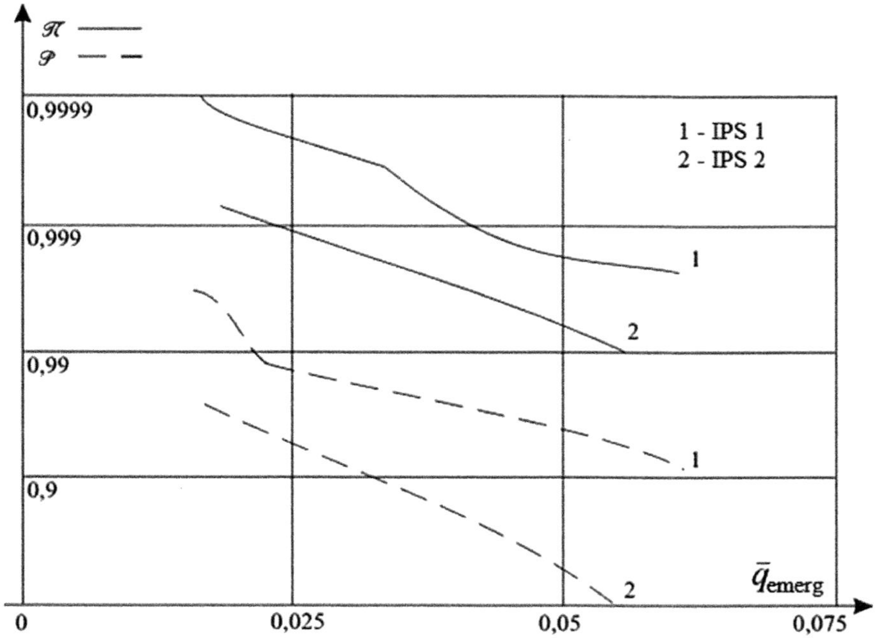

Fig. 4.5 $\pi = f_1(\bar{q}_{emerg})$ and $\mathscr{P} = f_2(\bar{q}_{emerg})$ for real systems

real conditions of equipment running-in $\bar{q}_{emerg\,run}$ during the forecasted growth of installed capacity of the researched IPSs.

"Increase in the accident rate due to new equipment commissioning" K^{emerg}_{commis}, shown in Table D.12, clearly illustrates the degree of decrease in the reliability of the equipment due to the running-in factor.

A higher emergency rate of equipment during the running-in period can increase the expectation of the capacity under emergency outage on the average by 50–70% (and in some cases above 100–120%) depending on the commissioning rates of new and, particularly, large-modular capacity. All these facts testify the necessity of taking account of system equipment running-in in the reliability calculations to obtain more accurate results. The above said is also correct for the rising emergency rate with the equipment aging.

Figure 4.6 shows the dependence $r_{op} = f(\bar{q}_{emerg})$ for an idealized system. The same dependences are typical of real systems (IPSs) at different values of π (Fig. 4.7).

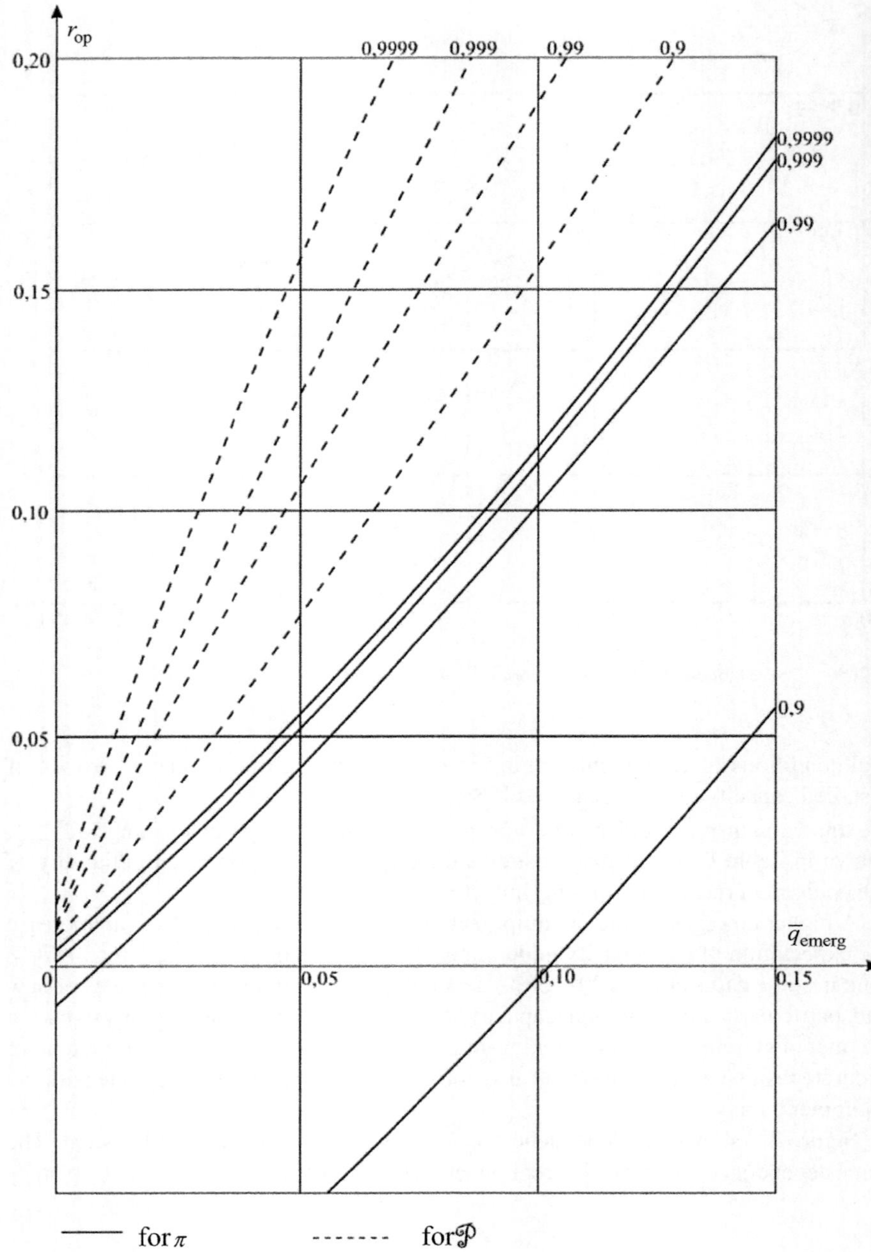

Fig. 4.6 Dependences of $r_{op} = f(\bar{q}_{emerg})$ with $\mathscr{P} = \text{const}$

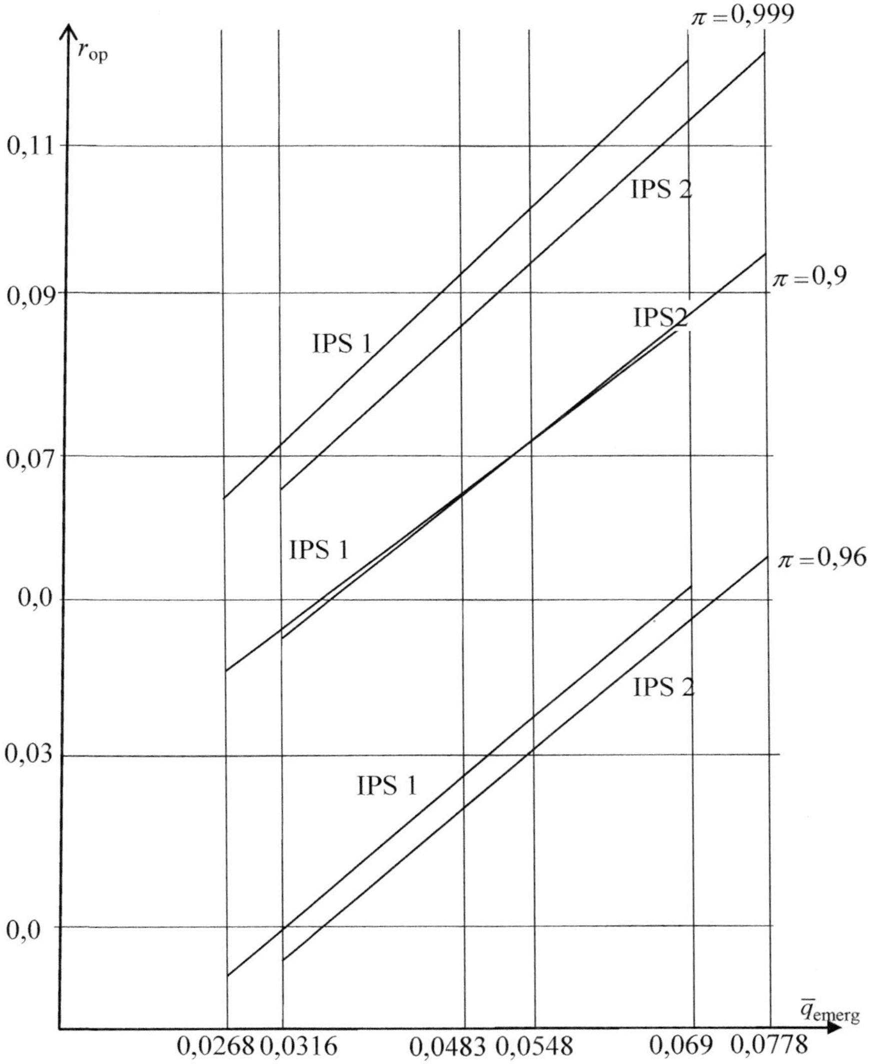

Fig. 4.7 Dependences of r_{op} with $\mathscr{P}$ or π = const (the initial data are given in Appendix D)

From Figure, it follows that at the preset (normative) reliability index π, the required value of the emergency reserve is directly proportional to the generalized index of the equipment emergency rate $\bar{q}_{\text{emerg}}$. This fact is of practical significance, since it makes it possible to estimate the required value of $\bar{q}_{\text{emerg}}$ based on two calculated points. The latter can substantially reduce the necessary number of calculations for the analysis and adjustment of some EPS development variant.

Figure 4.8 presents the reliability indices as a function of the operating reserve r_{op} for an idealized system. The schedule shape for real systems is preserved.

The analysis of these dependences and especially the dependences for real systems allow the conclusion that for normal power system operation with a sufficiently high reliability ($\mathscr{P}$ above 0.99–0.999) the actual values of r_{op} should be at a level of 8–10%, and sometimes 15% (at high $\bar{q}_{\text{emerg}}$). As is known, not so long ago (40–50 years ago) in the majority of domestic systems, $r_{\text{оп}}$ was much lower than 8–10%, which explains in some cases insufficiently high reliability of their work, functioning at a reduced frequency for extended periods.

It also seems interesting to assess the impact degree of variations in the values of operation reliability indices for individual groups of equipment in the system. The calculation results of power supply reliability indices with change in the values of q for individual groups of units by ±5 and 10% are presented in Table 4.8. As is seen from Table, this fact influences the calculated value of W_{und}.

Table 4.8 demonstrates the following regularities:

(a) the almost linear dependence of ΔW_{und} on Δq is true, in this case

$$\Delta W_{\text{und}\,j} \approx k \cdot h_j \cdot \Delta q_j,$$

where k is the coefficient taking the values in the range of 2.0–3.0, as follows from the data obtained for different equipment compositions in different power systems;

h_j is the share of this j-th group of units in the mathematical expectation of EPS capacity under emergency outage:

$$h_j = P_j n_j q_j / \sum_{i=1}^{I} P_i n_i q_i;$$

(b) the degree of the impact of change in q_j is the higher, than greater is the share of h_j. For the groups of units, the share of h in which does not exceed 30–50%, the change of q to ±10% leads to the change of ΔW_{und} less than by ±10%. Hence, the practical conclusion is that the values of q for the groups of units, whose share in the mathematical expectation of the capacity under emergency outage is higher than 30–50%, should be assigned most carefully. It is noteworthy that the share of h_j does not always correspond to the relative value of q_j, or to the value of the rated capacity P_j, or to the share of the total capacity $P_j n_j$ of the considered group in the total system capacity;

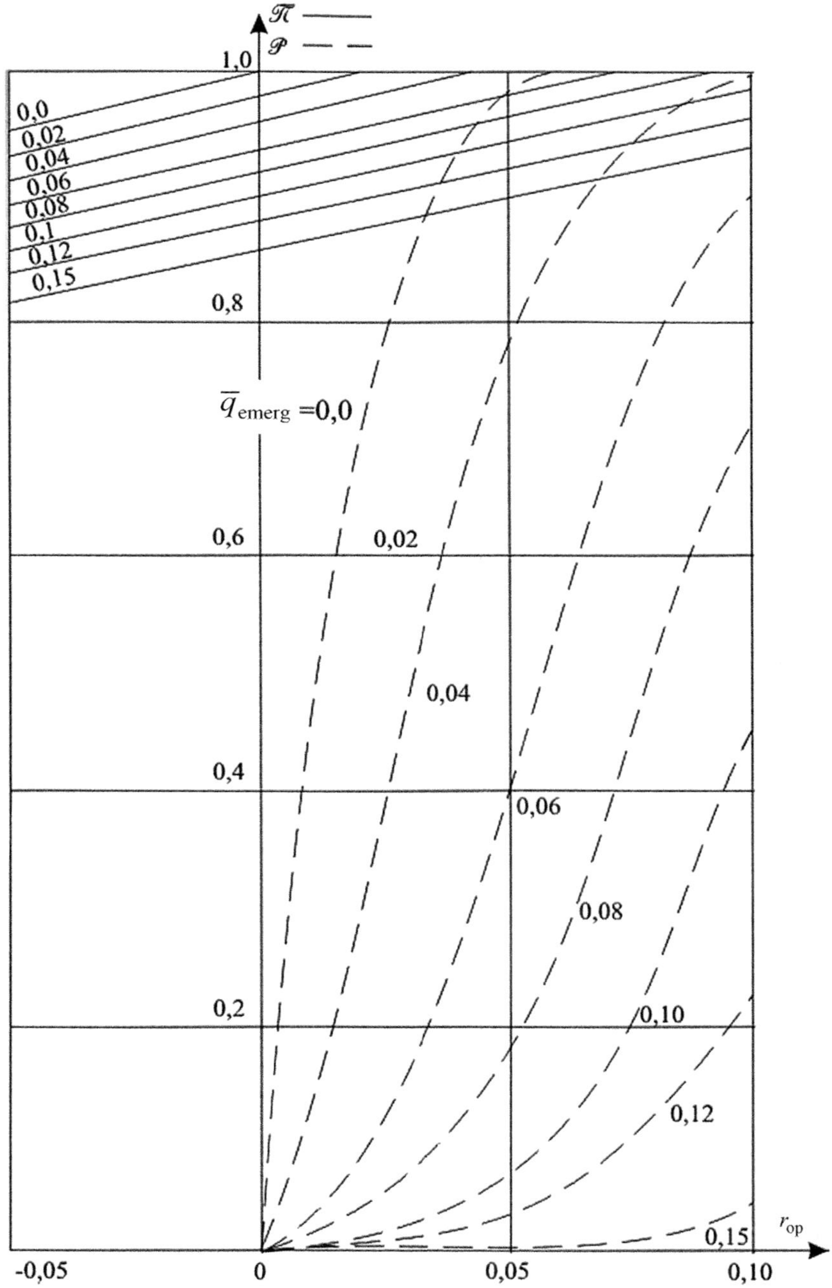

Fig. 4.8 $\pi = f_1(r_{op})$ and $\mathscr{P} = f_2(r_{op})$ for $\bar{q}_{emerg} = const$

Table 4.8 Estimation results of deviations in the calculated values of power supply reliability indices at the corresponding deviations of q of the units in individual groups

Δq		−10%		−5%		+5%		+10%	
Indices		RI	$\Delta\overline{RI}$, %	RI	ΔRI, %	RI	$\Delta\overline{RI}$, %	RI	$\Delta\overline{RI}$, %
IPS 3 $W_{und\,0} = 401.2$; $\mathscr{P}_0 = 0.90957$; $\pi_0 = 0.99811$									
1. Change of group with minimum, $h_j = 0.00242$	W_{und}	401	−0.05	401,08	−0,03	401,34	+0,03	401,24	+0,05
	$\mathscr{P}$	0.90960	+0.003	0.90959	+0.002	0.90955	−0.002	0.90953	−0.004
	π	0.99811	0	0.99811	0	0.99811	0	0.99811	0
2. Same of group with $h_j = 0.1645$	$W_{нед}$	387.94	−3.3	394.56	−1.7	407.91	+1.7	414.66	+3.36
	$\mathscr{P}$	0.91191	+0.25	0.91069	+0.12	0.90844	−0.12	0.90731	−0.25
	π	0.99817	+0.006	0.99814	+0.003	0.99808	−0.003	0.99804	−0.007
3. Same of group with maximum $h_j = 0.366$	W_{und}	368.68	−8.1	384.52	−4.2	418.18	+4.2	444.00	+10.7
	$\mathscr{P}$	0.91470	+0.56	0.91218	+0.29	0.90696	−0.29	0.90376	−0.64
	π	0.99826	+0.015	0.99819	+0.008	0.99803	−0.008	0.99790	−0.021
IPS 7 $W_{und\,0} = 32.022$; $\mathscr{P}_0 = 0.98129$; $\pi_0 = 0.99952$									
1. Change of group with minimum $h_j = 0.0122$	W_{und}	31.931	−0.29	31.968	−0.17	32.077	+0.17	32.113	+0.29
	$\mathscr{P}$	0.98135	+0.006	0.98132	+0.003	0.98125	−0.004	0.98123	−0.006
	π	0.99952	0	0.99952	0	0.99952	0	0.99952	0
2. Same of group with $h_j = 0.286$	W_{und}	318.07	−0.67	319.14	−0.34	321.13	+0.34	32.259	+0.68
	$\mathscr{P}$	0.98140	+0.012	0.98134	+0.005	0.98123	−0.006	0.98117	−0.012
	π	0.99952	0	0.99952	0	0.99952	0	0.99952	0
3. Same of group with maximum $h_j = 0.68$	W_{und}	25.357	−20.8	28.511	−11.0	35.784	+12.0	39.655	+23.8
	$\mathscr{P}$	0.98451	+0.33	0.98296	+0.17	0.97953	−0.18	0.97777	−0.35
	π	0.99962	+0.01	0.99957	+0.005	0.99946	−0.006	0.99940	−0.01

(c) by virtue of the interrelation among IPSs, the factual shares of h_j of each group of generating units at individual energy nodes in complex power systems in the mathematical expectation of the capacity under emergency outage in the whole system decreases considerably and hence, the values of $\Delta W_{\mathrm{und\,j}}$ also decrease because of the deviation of Δq_j. The deviation is the greater, the larger is the power system capacity, and the better is the development of its transmission part.

The analysis of the factors characterizing the compositions, structure, and parameters of generating equipment leads us to the following most important conclusions.

1. All the considered factors, namely the number of installed units, their rated capacities, and the emergency rates have an essential effect on the calculated levels of power system reliability and therefore, their correct accounting in the reliability models is necessary.
2. If possible, it is important to take account of all conditions determining unreliability of equipment operation: running-in, deviations of the emergency rate from the statistically average due to climatic and other conditions on the site of its installation, etc., since the described dependences vividly illustrate the impact of the assigned unavailability levels of generating equipment on the sought system reliability indices.
3. Of the total composition of generating capacity in EPS, the parameters of those groups of units have the greatest impact on the level of reliability of power supply, the share of which in the mathematical expectation of the total emergency power of the EPS is greatest. Hence, the system reliability can correctly be assessed first of all by the most thorough setting of parameters specifically for these groups of equipment.

4.2.3 Composition, Structure, and Parameters of Backbone Networks

The quantitative estimates of the impact of transfer capabilities of transmission lines and their emergency rates on the sought reliability indices are obtained first of all on the basis of the performed researches (see Tables 4.9 and 4.10).

Table 4.9 shows the role of the transfer capability values of transmission lines in maintaining reliable operation of power systems. For example, for the preset conditions of change in transfer capabilities of transmission lines, the undersupply can vary by more than 20–300 times for individual IPSs.

Table 4.9 Calculation results of reliability indices for different IPSs with different transfer capabilities of interconnection ties

IPS	Indices	Variant I $P_{in} = 0$		Variant II (actual)	Variant III $P_{in} = 5000$	
		RI	$\Delta\overline{\text{RI}}$, %	RI	RI	$\Delta\overline{\text{RI}}$, %
1	W_{und}	3.3577	−84	21.35	4.23	−80
	$\mathscr{P}$	0.99889	+0.24	0.99660	0.99540	- 0.11
	π	0.99998	+0.002	0.99996	0.999975	+0.001
2	W_{und}	769.1	+992	70.4	7.6655	−831
	$\mathscr{P}$	0.89365	−8.85	0.98045	0.99540	+1.5
	π	0.83735	+0.26	0.99472	0.999974	+0.5
3	W_{und}	6.274	+16.4	5.39	2.7137	−48
	$\mathscr{P}$	0.99476	+0.02	0.99454	0.99540	+0.09
	π	0.99994	−0.001	0.99995	0.999975	+0.002
4	W_{und}	384.23	+560	58.26	5.6108	−91
	$\mathscr{P}$	0.91084	−7.5	0.98486	0.99540	+1
	π	0.99819	−0.13	0.99945	0.999974	+0.05
5	W_{und}	1844.7	+558	280.25	8.9417	−96.8
	$\mathscr{P}$	0.82576	−13	0.94968	0.99539	+4.6
	π	0.99456	−0.46	0.99915	0.999974	+0.08
6	W_{und}	31.626	+30,310	0.104	1.6078	−1542
	$\mathscr{P}$	0.98152	−1.84	0.99994	0.99540	−0.45
	π	0.99952	−0.05	0.999999	0.999976	−0.003
7	W_{und}	0.4×10^{-7}	−98	0.24×10^{-5}	1.3217	–
	$\mathscr{P}$	0.99999	−0.0001	0.999999	0.99540	−0.46
	π	0.999999	0	0.999999	0.999976	–
8	W_{und}	4.7926	−8.5	5.24	1.6588	−70
	$\mathscr{P}$	0.99697	+0.3	0.99395	0.99540	+0.14
	π	0.99992	+0.001	0.99991	0.999973	+0.006
UPS	W_{und}	3044.08	+590	441	33.75	−92
	$\mathscr{P}$	0.997669	−0.2	0.99966	0.999974	+0.03

For UPS as a whole, the deviations of variant II from variants I and III amount to +590 and −92%, respectively. To the point, for the researches UPS scheme the calculated undersupply in the whole system is 32.07 million kW h, $\mathscr{P} = 0.99541$, $\pi = 0.999375$ on the assumption of the "unlimited" transfer capabilities of "absolutely reliable" interconnection ties (in comparison with the similar values of variant III in Table 4.9).

Table 4.10 Calculation results of reliability indices for different IPSs with different emergency rates of transmission lines

IPS	Indices	Variant A $q_{tie} = 0$		Variant B (actual)	Variant C $q_{tie} = 10q_{tie\ actual}$	
		RI	$\Delta\overline{RI}$, %	RI	RI	$\Delta\overline{\overline{RI}}$, %
1	W_{und}	4.2302	+0.005	4.2300	3.6485	−13.7
	$\mathscr{P}$	0.99541	+0.001	0.99540	0.99680	+0.14
	π	0.999975	0	0.999975	0.999978	0
2	W_{und}	7.3616	−3.96	7.6655	11.9582	+56
	$\mathscr{P}$	0.99541	+0.001	0.99540	0.98900	−0.64
	π	0.999975	0	0.999974	0.999959	−0.002
3	W_{und}	2.7594	+1.68	2.7137	1.9887	−26.7
	$\mathscr{P}$	0.99541	+0.001	0.99540	0.99571	+0.03
	π	0.999975	0	0.999975	0.999982	−0.001
4	W_{und}	5.3730	−4.24	5.6108	6.5783	+17.2
	$\mathscr{P}$	0.99541	+0.001	0.99540	0.99537	−0.003
	π	0.999975	0	0.999974	0.999969	−0.0005
5	W_{und}	8.6009	−3.81	8.9417	17.9442	+100
	$\mathscr{P}$	0.99541	+0.001	0.99539	0.98459	−1.08
	π	0.999975	0	0.999974	0.999947	−0.003
6	W_{und}	1.6877	+4.97	1.6078	0.9312	−42
	$\mathscr{P}$	0.99541	+0.001	0.99540	0.99946	+0.4
	π	0.999975	0	0.999976	0.999986	+0.001
7	W_{und}	1.4101	+6.7	1.3217	0.20201	−85
	$\mathscr{P}$	0.99541	+0.001	0.99540	0.99991	+0.45
	π	0.999975	0	0.999976	0.999997	+0.002
8	W_{und}	1.5434	−6.96	1.6588	1.8089	+9.0
	$\mathscr{P}$	0.99541	+0.001	0.99540	0.99533	−0.007
	π	0.999975	0	0.999973	0.999970	−0.0003
UPS	W_{und}	32.97	−2.3	33.75	45.06	+33.5
	$\mathscr{P}$	0.999975	+0.001	0.999974	0.999965	−0.001

It is noted that the values of the indices $\mathscr{P}$ and π are equalized for individual IPSs with the increasing transfer capabilities of ties (if capacity deficit is distributed proportionally to the IPS loads). For example, from Table 4.9 for different variants, the following values of these indices are:

Variant I	$\mathscr{P}_{\min} = 0.82576$	$\mathscr{P}_{\max} = 0.999999$
	$\pi_{\min} = 0.99456$	$=\pi_{\max} = 0.999999$
Variant II	$\mathscr{P}_{\min} = 0.94968$	$\mathscr{P}_{\max} = 0.999999$
	$\pi_{\min} = 0.99915$	$\pi_{\max} = 0.999990$
Variant III	$\mathscr{P}_{\min} = 0.99539$	$\mathscr{P}_{\max} = 0.99540$
	$\pi_{\min} = 0.999976$	$\pi_{\max} = 0.999976$

Comparison of $\mathscr{P}$ and π for the "concentrated" UPS reveals that the network of Variant III practically satisfies the available installed generating capacity of the power system.

From Table 4.10 that the role of the existing, relatively small accident rate of the transmission line is negligible (comparison of options A and B). However, with an increase in the accident rate by an order of magnitude for the cases of one or two powerful lines in communication, the failure of the transmission line may be more significant (comparison of variants B and C in Table 4.10).

It is easy to show in a similar way that the share of W_{und}, which depends on the structure and parameters of transmission lines, increases with decrease of the emergency rate of generating equipment and with increase of the system reserves, i.e., virtually, the natural and obvious relations between the factors determining power system operation reliability are confirmed.

Thus, with respect to the composition, structure, and parameters of backbone networks, the conclusions are as follows:

1. Transfer capabilities of interconnection ties are an essential factor determining the power system reliability level and therefore, they should be taken into account in the reliability models in the most complete and exact form, if possible.
2. The calculations show that the impact of the emergency rate of transmission lines on the sought reliability is negligible and therefore, sometimes this factor may be neglected. However, in the case of a longer outage of the transmission line (in some seasons of the year) or a special operation condition (the limiting one-sided load of a single transmission line), etc., the engineering assessment of such possibility is always needed.

4.3 Analysis of the Impact of Forms and Accuracy of Modeling the Power System Operation and Development Conditions on the Calculated Reliability Indices

The influence of the ways of presenting the initial information on the calculated (sought) indicators of reliability is discussed below. Here we will only list this initial information: power consumption schedules, generation and ties; random load fluctuations; accuracy of representation of the series of distribution of states of generating capacity; values of steps for calculating the series of distribution of states of generating capacity.

The researches of the required forms and accuracy of the initial data set in the proposed computational model were carried out based on the variation of a certain investigated factor with the values of all the others corresponding to their values in calculations, referred to below as "reference".

The reference calculations of power system reliability were made on the basis of sufficiently complete and exact representation of the above factors: daily load schedules were set by all 24 h; the normal law of distribution of random load fluctuations was approximated by seven stages; the generating equipment was represented by real groups and parameters; the calculated capacity step was taken equal to the multiple of the rated capacities of units; the order of magnitude of the probabilities assumed in calculations was 10^{-17}; the repairs were analyzed in the form presented in Appendix A; the outage probabilities were considered for any number of transmission lines.

During calculations, each studied factor was represented in a simplified form. The deviation of the obtained reliability indices from the reference values by no more than 5–10% (the generally accepted accuracy of engineering calculations) was considered acceptable for using this degree of simplification in power system reliability calculations. Difference in the values of the sought indices above 5–10% was the criterion of inadmissible application of this form or degree of simplification of the factor in calculations.

The analysis results of the impact of forms and accuracy of modeling the factors on the calculated reliability indices that are presented in this Section are based on the mass calculations of both multi- and single-nodal schemes of power systems. The results of reliability calculations of the multi-nodal schemes and in some cases the results for the single-nodal schemes, since they are quite sufficient for the performed analysis, are presented as examples. In this case, the visualization of the results increases without loss of their generality.

4.3.1 Approaches for Load Representation

In some known approaches to power system reliability assessment, the load schedules are represented as a load duration schedule or by a normal distribution function. In some works, they are represented as a linear section replacing a real load schedule or even as a constant value. At the same time in other works, the forms of representing power consumption schedules in the calculation model are showed to have an essential impact and therefore, it is recommended to use full typical daily power consumption schedules. Naturally, this is particularly expedient for multi-nodal models, which should take into account the time zones and the hourly correlation of load of different nodes.

The calculation results in Sect. 4.2.1 and in the work [5] show that if the real load schedules are replaced by the level of the yearly peak load, the error in determination of W_{und} reaches 15–25%. However, the calculations also show that for the used real load schedules the analysis of 4–10 h of the highest loads is admissible. The error in this case does not exceed 5%. But it is obvious that the range of analyzed loads depends on the density of load schedule: the higher is the density, the greater is the amount of the analyzed hours. The loads above (0.8/0.85) × P^1_{max}[2] have a meaningful impact on the value of reliability indices. The results obtained correlate well with the data presented in [1].

The performed researches allow the conclusion that the typical daily schedules are a more feasible form of setting the initial data on loads. In this case, the reliability of serving the hourly loads, starting with the maximum ones, is assessed in the model. The reliability calculation can terminate at the load values, for which, for example, the calculated value of undersupply or some other index is not higher than the specified value. In the "KORALL" and "YANTAR" programs, this fact is taken into account automatically in the block of screening the deficit-free states.

4.3.2 *Approaches for Accounting of Random Load Fluctuations*

From the contents of Sect. 4.2.1, it follows that the random load fluctuations are an essential factor that determines a power supply reliability level. In our opinion, however, the methods for their accounting have been devised insufficiently. The applied approach for accounting of this factor by the full normal law should apparently be adjusted, since the distribution truncation level of random fluctuations is evident due to the impossibility of load to be infinite.

The solution to this problem is beyond the scope of this work and therefore, it deals only with the traditional approaches for accounting of random load fluctuations.

A pronounced impact of random load fluctuations on the power system reliability level is noted in some works presenting quantitative characteristics of such an impact. They also indicate that the simplified consideration of this factor using the absolute (so-called irregular) yearly peak load leads to reliability underestimation. The use of regular load values (without accounting of load fluctuations) results in overestimation.

The "KORALL" program was applied to test the following forms of accounting of random load fluctuations:

(1) assignment of the normal law of random fluctuations by five stages of the approximating schedule (range $\pm 2.5\sigma$; probabilities 0.07; 0,24; 0.38; 0.25; 0.07);
(2) the same for three stages (range $\pm 3\sigma$; probabilities 0.0668; 0.8664; 0.0668);
(3) application of the irregular load values calculated by the formula in [4] in the calculations (instead of the regular load values and the approximating schedule of its random fluctuations)

[2]When the scheduled repairs of generating facilities are taken into account in the conventional load schedule (see Appendix A.4), the factually accounted loads are equal to (0.75/0.8) P^1_{max}.

$$P^{1}_{irreg} = P^{1}_{max} + \alpha,$$

where $\alpha = \alpha' = 0.01\,P^{1}_{max} + 1,26\sqrt{P^{1}_{max}}$;

(4) the same, but $\alpha = \alpha'' = 1,645 \cdot \sigma \cdot P^{1}_{max}$;

(5) the application of the regular load values solely without random fluctuations.

The calculations show (see Table 4.11) that the use of the five-stage schedule delivers almost the same results as the seven-stage one. The use of three stages leads in some cases to the error in the assessment of W_{und} above 10%.

The accounting of random fluctuations by forms 3 and 4 (in this case $\alpha'' > \alpha'$) results in the essential overestimation of W_{und}.

Neglect of random load fluctuations (approach 5) leads, in general, to the underestimation of W_{und} (correspondingly, of other indices) up to 20–70% in comparison with the reference calculation.

From the above said, it follows that:

(a) five stages are sufficient for approximation of the normal distribution law of random load fluctuations in practical calculations;

(b) for accounting of random load fluctuations with the help of approaches 3 and 4, the calculated values of α should be adjusted.

4.3.3 Approaches for Generating Capacity Representation

This Section is devoted to the researches of different methods of equivalent representation of generating equipment in computational models (see, for example, [6]):

(1) the equivalenting in terms of the equality of the total capacities of units, the mathematical expectation of the values of capacities under emergency outage and the mathematical expectation of durations of deficit-free system state in the equivalent (homogeneous) and real (heterogeneous) generating systems. The equivalenting formulas obtained from these conditions have the following form:

$$P_{equival} = \frac{\sum_{i=1}^{n} P_i}{n}; \quad n_{equival} = \frac{\sum_{i=1}^{n} \lg(1-q_i)}{\lg(1-q)}; \quad q_{equival} = \frac{\sum_{i=1}^{n} q_i P_i}{\sum_{i=1}^{n} P_i} \tag{4.5}$$

where $P_{equival}$ is the capacity of the equivalent unit; $n_{equival}$ is the amount of the equivalent units; $q_{equival}$ is the emergency rate of the equivalent units;

Table 4.11 Calculation results of reliability indices (RI) for some power systems by different approaches of accounting of random load fluctuations

Reliability indices	Normal law, 7 stages	Normal law, 5 stages		Normal law, 3 stages		$P^{1}_{\text{irreg}\,1}$		$P^{1}_{\text{irreg}\,2}$		Without random fluctuations	
	RI (reference)	RI	$\Delta\overline{\text{RI}}$, %	RI	$\Delta\overline{\text{RI}}$, %	RI	$\Delta\overline{\text{RI}}$, %	RI	$\Delta\overline{\text{RI}}$, %	RI	$\Delta\overline{\text{RI}}$, %
IPS 5$\sigma = 0.022$											
W_{und}	2202.4	2194.1	−0.38	2020.7	−0.25	3098.1	+40.7	8262.7	+275	1844.7	−16.2
$\mathscr{P}$	0.82241	0.82263	+0.03	0.82393	+0.19	0.76209	−7.34	0.60184	−27	0.82476	−0.41
π	0.99351	0.99353	0.00	0.99404	+0.05	0.99101	−0.25	0.97702	−1.7	0.99456	+0.11
IPS 1$\sigma = 0.031$											
W_{und}	13.711	12.958	−8.41	9.032	−34.1	9.379	−31.6	54.585	+298	3.350	−75.5
$\mathscr{P}$	0.99636	0.99636	+0.02	0.99752	+0.12	0.99601	+0.06	0.98530	−0.02	0.99998	+0.01
π	0.99992	0.99992	0	0.99995	+0.003	0.99994	+0.002	0.99969	−0.02	0.99998	+0.01
IPS 6 $\sigma = 0.023$											
W_{und}	124.56	120.53	−3.19	81.493	−34.6	85.428	−30.6	536.50	+331	31.626	−74.6
$\mathscr{P}$	0.95227	0.95243	+0.017	0.96782	+1.63	0.95518	+0.31	0.81073	−14.9	0.98152	+3.07
π	0.99813	0.99819	+0.006	0.99878	+0.07	0.99873	+0.06	0.99253	−0.56	0.99952	+0.14

(2) replacement of the last condition in the first approach by the equality condition of variances of the capacity under emergency outage results in the following formula for $n_{equival}$ [provided that $p_{equival}$ and $q_{equival}$ are the same as in (4.5)]:

$$n_{equival} = \sum_{i=1}^{n} q_i P_i \left(\sum_{i=1}^{n} P_i - \sum_{i=1}^{n} P_i q_i \right) / \sum_{i=1}^{n} q_i (1 - q_i) P_i^2; \tag{4.6}$$

(1) with $p_{equival}$ and $q_{equival}$ corresponding to (4.5) it appears that

$$n_{equival} = n; \tag{4.7}$$

(2) with $p_{equival}$ and $q_{equival}$ corresponding to (4.6) it appears that

$$n_{equival} = \left(\sum_{i=1}^{n} q_i P_i \right)^2 / \sum_{i=1}^{n} P_i^2; \tag{4.8}$$

(3) for determination of the parameters of a homogeneous system of equivalent units, the equivalenting in terms of the equality of the first three ways for probability distribution of emergency capacity loss for the equivalent and real generation systems results in the following expressions

$$P_{equival} = \frac{2M_2^2 - M_1 M_3}{M_1 M_2}; \; n_{equival} = \frac{M_1^2 M_2}{2M_2^2 - M_1 M_3}; \; q_{equival} = \frac{M_2^2 - M_1 M_3}{2M_2^2 - M_1 M_3}, \tag{4.9}$$

where the corresponding methods are equal to:

$$M_1 = \sum_{i=1}^{n} q_i P_i; \; M_2 = \sum_{i=1}^{n} q_i (1 - q_i) P_i^2; \; M_3 = \sum_{i=1}^{n} q_i (1 - q_i)(1 - 2q_i) P_i^3.$$

The examples of calculations of the reliability indices using different equivalenting methods[3] and their comparison with the reference calculations (with the step multiple of the rated capacities of generating units in real groups) are demonstrated in Table 4.12.

Table shows that in the majority of cases the equivalenting methods formally provide an acceptable calculation accuracy of $\mathscr{P}$ and π, nevertheless give unsatisfactory results at assessment of W_{und} (the error considerably exceeds the admissible 5–10%). Note the instability of the error sign in reliability calculations of different systems, which causes high doubt of using these equivalenting methods even for comparative analysis of the variants.

---

[3]It was test impossible to method 5, as far as in this case the values of $n_{equival}$ and $q_{equival}$ are negative. For example, for one of IPSs $p_{equival} = 280$; $n_э = -15.29$; $q_{equival} = -0.497$.

Table 4.12 Calculation results of the reliability indices for some power systems by different methods of equivalenting of generating units in the calculation model

Equivalenting methods	Reference	I		II		III		IV	
	RI	RI	$\Delta\overline{\text{RI}}$, %	RI	$\Delta\overline{\text{RI}}$, %	RI	$\Delta\overline{\text{RI}}$, %	RI	$\Delta\overline{\text{RI}}$, %
IPS 5 $q_{equival} = 0.048$									
Parameters of equivalent group	–	$P_{\text{э}} = 1267;\ n_{\text{э}} = 35$		$P_{\text{э}} = 443;\ n_{\text{э}} = 101$		$P_{\text{э}} = 162;\ n_{\text{э}} = 273$		$P_{\text{э}} = 343;\ n_{\text{э}} = 128$	
W_{und}	1844.7	2513.6	+36.6	1307.7	−29.1	1587.9	−13.9	1757.7	−4.52
$\mathscr{P}$	0.82576	0.82447	−0.16	0.81153	+1.98	0.82419	−0.19	0.82211	−0.44
π	0.99456	0.99257	−0.2	0.99615	+0.16	0.99532	+0.08	0.99483	+0.03
IPS 1 $q_{equival} = 0.0769$									
Parameters of equivalent group	–	$P_{\text{э}} = 372;\ n_{\text{э}} = 65$		$P_{\text{э}} = 318;\ n_{\text{э}} = 76$		$P_{\text{э}} = 106;\ n_{\text{э}} = 227$		$P_{\text{э}} = 306;\ n_{\text{э}} = 79$	
W_{und}	3.258	53.085	+1481	38.153	+1036	1.343	−59.9	32.544	+869
$\mathscr{P}$	0.99889	0.98635	−1.26	0.98320	−1.07	0.99904	+0.02	0.98960	−0.93
π	0.99968	0.99998	−0.03	0.99977	−0.02	0.99999	+0.001	0.99880	−0.93
IPS 6 $q_{equival} = 0.3676$									
Parameters of equivalent group	–	$P_{\text{э}} = 176, 4;\ n_{\text{э}} = 53$		$P_{\text{э}} = 240;\ n_{\text{э}} = 39$		$P_{\text{э}} = 123;\ n_{\text{э}} = 76$		$P_{\text{э}} = 303;\ n_{\text{э}} = 46$	
W_{und}	31.626	17.883	−43.5	29.927	−5.37	8.851	−72	22.23	−29.7
$\mathscr{P}$	0.98152	0.98564	+0.04	0.97956	−0.2	0.98989	+0.85	0.97997	−0.18
π	0.99952	0.99973	+0.02	0.99955	+0.002	0.99987	+0.04	0.99967	+0.02

4.3.4 Exponent of Probabilities Considered in Calculations

In all the known recommendations regarding the necessary order of the considered probabilities of the states of the generating equipment in the nodes, when performing reliability calculations, it is suggested to take into account probabilities of order not less than 10^{-5}–10^{-6} [7, 4, etc.].

When determining the reliability indices for the systems considered in this work, it was revealed that the analysis of only the probabilities with the exponent of no less than 10^{-5} offered practically the same results as in the reference calculation that considered probabilities of up to 10^{-17}. Thereby, the computing time naturally decreased. In separate cases, it turned out sufficient to consider the probabilities with the exponent of 10^{-3} (see Table 4.13).

Thus, as concerns the exponent of the probabilities considered in calculations, it can be concluded that for practical reliability calculations of real power systems, it is sufficient to consider the states with the probabilities of no less than 10^{-3} to 10^{-4}. This fact allows the computing time saving without an essential degradation of accuracy.

4.3.5 Step for Calculation of Distribution Series of Generating Capacity

The reliability indices are calculated using the standard mathematical practice of replacing some real function (here—distribution function) by some discrete values with a preset step (see Fig. 4.9). Hence, to determine the step of ΔP, which is maximum in magnitude and admissible in terms of the accuracy of determining the reliability indices on the basis of discrete sought values, becomes necessary. The bigger is the step, the shorter is the computing time, and the lower are the memory requirements, but the accuracy of function representation is lower.

The maximum admissible step of distribution series of the capacity under emergency outage was estimated in the reliability calculations based on the performed research. Some results of the study are illustrated in Table 4.14.

Comparison of changes in the sought reliability indices (particularly W_{und}) with change in the step reveals that the maximum admissible step should be equal to $P_{\text{step}} = \Delta P \leq P^{\text{g}}_{\text{instal}}/(100 \div 500)$, if P_{step} in this case exceeds the rated capacities of generating units in the groups. Otherwise, P_{step} is taken equal to the rated capacities of the corresponding group of generating units. However, it should be noted that this dependence is determined first of all by the accepted algorithm for calculation of the distribution series of generating capacity at the node. The dependences presented in Table 4.14 are true for the algorithm described in Appendix A.5. As follows from the calculations, it is characterized by decrease in the sought reliability level with increasing step.

Table 4.13 Examples of change in the sought reliability indices of power systems as a function of the exponent of probabilities of equipment state that are analyzed in calculations

Reliability indices	10^{-17}	10^{-10}		10^{-5}		10^{-3}		10^{-2}	
	RI (reference)	RI	$\Delta\overline{RI}$, %	RI	$\Delta\overline{RI}$, %	RI	$\Delta\overline{RI}$, %	RI	$\Delta\overline{RI}$, %
IPS 5									
W_{und}	1844.7	1914.7	0	1844.7	0	1840.4	−0.23	1775.4	−3.75
$\mathscr{P}$	0.82576	0.82576	0	0.82576	0	0.82589	+0.016	0.82817	+0.03
π	0.99456	0.99456	0	0.99456	0	0.99458	+0.002	0.99477	+0.02
IPS 1									
W_{und}	3.358	3.358	0	3.341	−0.51	1.2803	−62	0	−100
$\mathscr{P}$	0.99889	0.99889	0	0.99889	0	0.99925	+0.04	1.0	+0.12
π	0.99998	0.99998	0	0.99998	0	0.99999	+0.001	1.0	+0.002
IPS 6									
W_{und}	31.625	31.625	0	31.608	−0.05	30.316	−4.14	19.133	−42.7
$\mathscr{P}$	0.98152	0.19852	0	0.98152	0	0.98168	+0.002	0.98364	+0.22
π	0.99952	0.99952	0	0.99953	0	0.99954	+0.002	0.99973	+0.02

Table 4.14 Examples of the sought reliability indices of power systems as a function of the value of P^{g}_{step} (MW and % of P^{g}_{avail})

IPS 6 $P^{g}_{avail} = 9350\ MW$											
Reliability indices	50 reference	100		200		3000		500		1000	
		1.07		2.14		3.21		5.25		10.7	
	RI	RI	$\Delta\overline{RI}$, %	RI	$\Delta\overline{RI}$, %	RI	$\Delta\overline{RI}$, %	R	$\Delta\overline{RI}$, %	RI	$\Delta\overline{RI}$, %
W_{und}	31,626	32.022	+1.25	38.814	+1634	39.319	+24.3	80.178	+153	202.93	+542
$\mathcal{P}$	0,98152	0.98129	−0.02	0.98058	−0.1	0.98064	−0.09	0.97222	−0.94	0.93866	−4.3
π	0,99952	0,99952	0	0.99945	−0.01	0.99941	−0.011	0.99880	−0.07	0.99695	−0.26
UPS $P^{g}_{avail} = 178{,}300\ MW$											
Reliability indices	50 reference	250		500		1000		5000		10,000	
		0.14		0.28		0.56		2.3		5.6	
	RI	RI	$\Delta\overline{RI}$, %	RI	$\Delta\overline{RI}$, %	RI	$\Delta\overline{RI}$, %	RI	$\Delta\overline{RI}$, %	RI	$\Delta\overline{RI}$, %
W_{und}	33.870	34.025	+4.57	45.517	+34.4	77.298	+127	562.14	+1560	934.55	+2650
$\mathcal{P}$	0.99538	0.99537	−0.001	0.99484	−0.054	0.99316	−0.22	0.97664	−1.05	0.97261	−2.28
π	0.99997	0.99997	0	0.99997	0	0.99994	−0.003	0.99957	−0.04	0.99928	−0.07

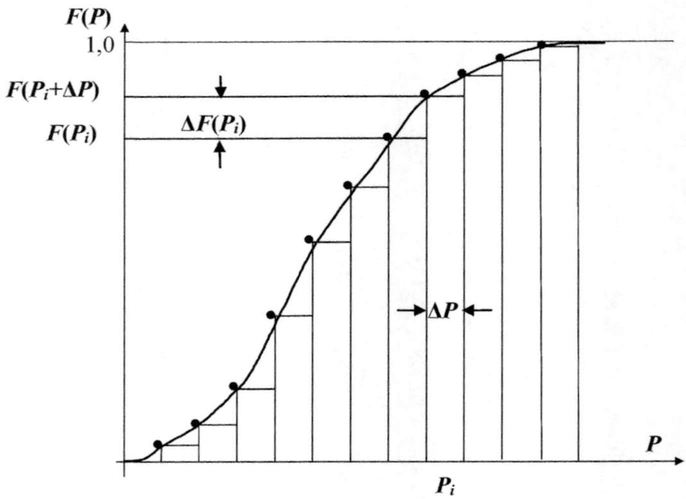

Fig. 4.9 Substitution of the distribution function for the discrete series of probability values

4.3.6 Approach for Representation of Ties

In contrast to generating equipment, the methods for representation of ties in the **calculation** reliability models known by publications are not diverse.

In principle, under the assumptions made, the methods may differ in:

(1) representation of transmission line outages under scheduled repairs;
(2) representation of ties with a complex connection scheme of transmission lines (with bypasses, links, switching stations, and so on);
(3) analysis of the states of ties consisting of more than two transmission lines.

As for the outages of transmission lines under scheduled repairs, there is no real alternative to the method used in the "KORALL" and "YANTAR" program: the duration of scheduled repairs (in p.u.) at certain periods of a year is added to the durations of their emergency outages. As is known, in other models, both the scheduled and emergency outages are usually not taken into account for the reasons mentioned in Sect. 4.2.3.

The corresponding recommendations concerning the representation of ties with a complex connection scheme are described in [8].

As concerns the representation of distribution series of the states of ties containing more than two transmission lines, the practice revealed the following. Because of the relatively low values of q^{tl} (as a rule from 0.001 to 0.01/100 km), the indicated distribution series can include probabilities of the tie states, which are characterized by the outages of no more than 2–3 transmission lines. Actually, the probability of outages of more than two transmission lines usually proves to be much lower than 0.01×10^{-6}.

As is seen in Sect. 4.3.4, the states with probabilities of such and lower exponents do not practically influence the values of calculated reliability indices. In many cases with $q^{\mathrm{tl}} < 0.01$, it is sufficient for practical purposes to take into account only the states of ties, which are characterized by full operability and outages of one or two transmission lines.

* * *

Since in the course of the research, the form of setting or simplification was changed for each factor separately, after analysis of all of them, the calculations were made based on the initial data represented in their admissible simplified form.

The calculations were made with the following values of initial parameters: the normal law of load fluctuations random was represented by three stages; the value of $P^{\mathrm{g}}_{\mathrm{step}\,m}$ was taken equal to $P^{\mathrm{g}}_{\mathrm{avail}}/200$; the exponent of magnitude of the least probabilities was assumed to be 10^{-5}; the scheduled repairs of generating units were taken into account in the load schedule; the amount of transmission lines under simultaneous outage in each interconnection tie was taken equal to no more than two. Thereby, as was noted, in the program, the load ranking and the block of screening of deficit-free states enabled to terminate the calculations after analysis of the first 8–10 h of the highest loads.

These calculations for different variants of power system schemes show that the calculated values of reliability indices (including W_{und}) coincide with the reference indices with an error of up to 10%, and the duration of calculations decreases, on the average, fivefold.

Thus, the analysis performed showed that practically without reducing the accuracy of calculations, by rationally setting the initial data, it is possible to reduce the calculation time by the program and reduce the requirements for the accuracy of the initial data. Reduction of the counting time is especially effective when carrying out mass calculations or in conditions when the reliability calculation program is a sub-block repeatedly used in the optimization model of choosing the perspective structure of EPS.

References

1. Lyalik GN (1975) The role and importance of hydroelectric power stations in ensuring the optimal level of reliability of power supply for mixed energy systems and methods for their accounting in the design. In: Reports of the scientific and technical conference on the results of scientific research for 1963–1969 years. Electric Power Section. Subsection of hydropower. MEI, Moscow, pp 21–33 (in Russia)
2. Rozanov MN, Loginov MV (1967) Determination of the mathematical expectation of economic damage from the under–supply of electricity consumers in the unified energy system by the method of statistical tests. In: Proceedings of VNIIE, issue. XXX. Energia, Moscow, pp 84–97 (in Russia)
3. Rokotyan SS, Shapiro IM (1985) Handbook on the design of electric power systems. Energoatomizdat, Moscow, 352 p (in Russia)

4. Avramov NN, Bartosh L, Grabak J et al (1988) Reserves of capacity in the electric power systems of the CMEA member countries. Methods of research. Science. Siberian Branch, Novosibirsk, 150 p (in Russia)
5. Mogirev VV, Rudenko YN Algorithms for investigating the reliability of electric power systems of arbitrary configuration. Izvestiya AN SSSR. Energy Transp 1:38–45 (in Russia)
6. Rudenko YN, Cheltsov MB (1974) Reliability and redundancy in electro-energy systems. Methods of research. Science, Siberian Branch, Novosibirsk, 264 p (in Russia)
7. Volkov GA (1970) Calculation of the optimal reserve of active power in designing the development of power systems on the Ural–2 Computer. In: Problems of the optimal development of power systems and new technical means of protecting them. Nauka, Moscow, pp 3–12 (in Russia)
8. Aleksandrov IA, Kovalev GF, Rudenko YN (1973) Comparative analysis of reliability of various options for the development of electric power systems. In: Reports at the third all-union scientific and technical meeting on the stability and reliability of the energy systems of the USSR. Energia, Leningrad, pp 394–405 (in Russia)

Chapter 5
Investigation of the Reliability of Variants of the Perspective Structure of Power System

The approach described in Chap. 3 was applied to research different prospective variants of UPS development for different time periods. The results of the corresponding calculations and analysis are presented in [1 and others].

The reliability research of two variants of UPS development for the long run is presented in this paper as an illustration of the application of the developed approach.

Variant 1 is characterized by the development of intersystem connections during this period on the developed voltages of 330, 500, and 750 kV;
Variant 2 is characterized by the development of interconnection ties of 330, 500, 750, and 1150 kV with replacement of some transmission lines of 330, 500, 750 kV of the previous variant by the 1150 kV transmission lines.

5.1 Conditions of the Study and the Initial Data Characteristic

A special feature of the research is an attempt at a comprehensive analysis of the reliability of the EEC. It includes calculation of stationary admissible modes (calculated states); the estimation of the static stability of the main modes (calculated states) of the system; the estimation of dynamic stability under certain calculated perturbations; analysis of failure-free operation of electricity supply to consumers.

This chapter deals mainly with the analysis of reliable power supply to consumers.

The UPS reliability research should involve first of all reliability that is determined by the main structure of UPS: generating capacities of IPSs comprising it and interconnection ties. Therefore, the UPS scheme, in which IPSs (as nodes) and all interconnection ties among them are modeled according to the principles described in Chap. 3, is used for calculations. The parallel operation of six energy areas of the European part of UPS is considered in the research (the scheme of Appendix D.1).

1. The initial data are the results of the prospective designing of UPS that was carried on in due time by the Department of Energy System Designing at the

G. F. Kovalev and L. M. Lebedeva, *Reliability of Power Systems*, Power Systems, https://doi.org/10.1007/978-3-030-18736-1_5

Institute "Energosetproject" for the period up to 1990. The information required for calculations and analysis of the indicated variants is given in Appendix D.

The research is carried out for three calculated time intervals, which stand apart for five years.

The calculated electrical schemes of the EPS, as the results of equivalence, for different time levels are shown in Figs. D.2 and D.3.

In these schemes, each IPS to obtain the initial normal permissible states (modes) is represented by one equivalent generator and several nodes with adjacent intersystem ties. The internal ties are represented equivalently. Figures D.2 and D.3 show the results of the calculation of the electric modes for the period of the absolute maximum load for all intersystem transmission lines in operation. They also present the information about the lengths and commissioning time of transmission lines, their rated voltage and transfer capabilities.

The typical consumption schedules are illustrated in tables D.1–D.3. Two types of schedules: winter and summer are considered for each node. The daily schedules of IPSs are given in the local time.

Tables D.4–D.8 present the data on generating equipment, which is necessary for calculating power supply reliability. Depending on the equipment commissioning time, the projected probabilities of emergency outages are assumed with regard to the running-in period (from 1 to 3 years).

The standards of outage in scheduled repairs and the duration of emergency outages of power lines are shown in Tables D.9 and D.10.

Since the reliability of numerous variants of IPS development is assessed for the future, the relative durations of emergency outages of system elements for the considered periods are set as a range of possible values between the optimistic and pessimistic estimates. They are determined from the analysis of the published statistical data on power system equipment [2–7].

The average values of equipment reliability indices are calculated by the formula

$$q_{\text{average}} = \frac{q_{\text{opt}} + q_{\text{pessim}}}{2}.$$

The calculated values of q_{average} proved to be close to those applied at the Institute "Energosetproject" to similar researches.

5.2 Analysis of the Results of Researches

The reliability was calculated based on the data of UPS development. The calculation of power supply reliability indices characterizing power systems from different sides: $\mathscr{P}$, π and W_{und} is its specific feature.

The calculation results for the specified options of UPS development are presented in the tables and figures of Appendix D. The semilogarithmic scale is applied in the

Table 5.1 Main absolute and relative characteristics of UPS, which are equal for both considered variants at different calculated time intervals

Index	Units	Calculated intervals			
		Initial	1	2	3
Power generation	Billion kWh	743	1110	1589	2216
Installed capacity	Thousand MW	125	208	300	422
Installed capacity of modeled equipment	Thousand MW	37.5	80.8	146.9	249.2
Increase in power generation as against the initial (in numerator) and previous (in denominator) intervals	p.u.	$\frac{1,0}{-}$	$\frac{1,50}{1,50}$	$\frac{2,15}{1,43}$	$\frac{3,0}{1,39}$
Increase in installed capacity as against the initial (in numerator) and previous level (in denominator)	p.u.	$\frac{1,0}{-}$	$\frac{1,67}{1,67}$	$\frac{2,41}{1,45}$	$\frac{3,38}{1,40}$
Capacity of modeled equipment relative to total installed capacity of UPS	p.u.	0.30	0.39	0.49	0.59

plots, which allow a convenient representation of specific numerical values close to zero or unity and slightly differing from each other in the traditional representation.

The presented below text is devoted to the analysis of the results obtained. They enable to pay attention to some relations between the calculated values of reliability indices and specific features of the considered options of UPS development and also peculiarities in UPS operation and development from the viewpoint of reliability in general.

The results obtained (see figures in Appendices D.4–D.13 and tables in Appendices D.11 and D.12) allow a comparative analysis of UPS operation reliability for two variants of its network development. The researched options of UPS development are characterized by the high growth rates of installed capacity and generation, which is illustrated by the data in Table 5.1.

In such a case, the preferential commissioning of units of large rated capacities is the strategic trend. Thus, the share of large-block units in UPS rises from 0.30 at the initial interval to nearly 0.60 at the third time interval. As is known, the block equipment is characterized by the high economic indices, but sometimes has relatively low-security indices. As will be shown below, this fact has an essential impact on the level of UPS operation reliability.

As for the distinctions between two variants in the network structure, they are the following.

At the first calculated level, the 1150 kV transmission line is installed instead of one 500 kV transmission line between IPSs 2 and 3 (variant 1); one 1150 kV transmission line instead of two 500 kV transmission lines between IPSs 2 and 4 and also one 1150 kV transmission line instead of two 500 kV transmission lines between IPSs 4 and 6. The rest of the network in both options is the same (see the schemes in Appendices D.2 and D.3).

At the second calculated level, the variants differ in the following transmission lines:

- IPS 2–IPS 3. In the first variant, the 330 kV transmission line is added to the ties of the previous level; the scheme of the second variant remains at the previous level;
- IPS 3–IPS 4. In comparison with the previous level, three 500 kV transmission lines are added in the first variant and one 1150 kV transmission line in the second;
- IPS 4–IPS 6. As compared to the previous level, one 1150 kV transmission line is added in the first variant (see the schemes in Appendices D.2 and D.3).

At the second level, the distinctions in development variants as against the first level are as follows:

- The 750 kV transmission line between IPS 2 and IPS 3 is added in the first variant, and construction of new transmission lines is not planned in the second variant;
- Two 500 kV transmission lines are installed between IPS 2 and IPS 4 in the first variant, and one 1150 kV transmission line is installed in the second variant;
- One 500 kV transmission line is installed between IPS 4 and IPS 6 in the first variant, and construction of new transmission lines is not planned in the second variant.

The internal ties of IPSs in both variants develop in a similar way.

The Figures in Appendices D.4–D.6 present probable ranges of possible values of different reliability indices for IPSs and UPS, which were determined by the reliability calculations for the three calculated levels, specified conditions of systems development and possible ranges of change in the emergency rate of equipment.

The Figures in Appendices D.7–D.13 illustrate dynamics of change in the reliability indices over time from the first to the third levels of each IPS and UPS as a whole.

Analyzing the results presented on the graphs, we can state the following considerations.

1. The prospective options of UPS development are characterized by not very high-reliability indices. The calculated value of the index $\mathscr{P}$ is at a level of 0.98, and that of the index π—at a level of 0.999.
2. The calculation results show that the reliability of the second variant is higher than the reliability of the first variant at all three-time stages.

Comparison of the reliability indices of different IPSs in both options (figures in Appendices D.7–D.12) reveals that the reliability of the second variant is higher owing to the higher operation reliability of the largest IPSs within UPS: IPS 6 and particularly IPS 3 at practically the same operation reliability of IPS 4 and IPS 5 in both variants. However, the operation reliability of IPS 1 and IPS 2 in the second variant is somewhat lower than in the first variant at the first and second levels and for IPS 2—at the third level as well.

The key reason of the high-reliability level of the second variant as against the first consists in higher real transfer capabilities of the installed transmission lines of 1150 kV as compared to transfer capabilities of the transmission lines of 330, 500, and 750 kV of the first variant.

3. Despite the high reliability in general, it is seen that the prospective structure of UPS from the reliability standpoint was formed rather approximately. This fact became apparent from the random scatter of the ranges of probable values of reliability indices for different IPSs (see figures in Appendices D.4–D.6) and at different calculated time intervals (see figures in Appendices D.7–D.12).

The scatter of the values of reliability indicators is caused by the fact that indicators depend on many factors, which in design practice, due to complexity, are taken into account at the level of application of a number of mediated standards and engineering (expert) estimates.

The results show that the advantage of the suggested approach is the possibility of quantitative evaluation of operation conditions of individual IPSs within UPS in terms of reliability and the possibility of indication of the reasons determining some or other values of the reliability indices and the possible ways of changing these values as desired direction. The operation condition of IPSs (power consumption or its delivery over interconnection ties), the working schedule (load curves) of consumers, the structure of generating capacities, the dynamics of commissioning new equipment, the emergency rate of system elements, the reserve volume and allocation, the structure and transfer capabilities of ties with other IPSs are the factors determining the operation reliability level for each IPS. The calculations show that the relations between these factors for different IPSs vary to a great extent in the specific researched variants, which leads to diverse operation reliability levels of these IPSs in UPS.

The degree of impact on the factors determining operation reliability of the studied subsystems in UPS cannot be the same, as far as part of these factors objectively depends to a greater or lesser extent on the required external and internal strategic reasons for the development of EPS. They comprise, for example, power consumption schedules, degree of provision with own generating capacity (receiving or supplying subsystem), increase in the share of block capacity, level of equipment emergency rate. By contrast, it is possible to extensively influence other factors, for example, transmission capacities of ties, volume, and allocation of available emergency reserve to achieve the required reliability level of subsystems.

The position of each of six IPSs forming UPS is considered for illustration of the impact of different factors. The calculation results in figures of Appendices D.4–D.13, as well as the additional estimates in tables of Appendices D.11–D.12 are the information characterizing the position of each IPS from the standpoint of its reliable operation in UPS.

IPS 1. The table in Appendix D.11 shows that this IPS is characterized by the fact that its installed capacity makes up approximately a sixth part (0.17) of the capacity installed in UPS. The relative value of installed capacity of IPS 1 remains almost invariable at all calculated levels. The relative increases of capacity increases coincide with the average ones for UPS (see column 3 of table in Appendix D.11). The relative component of the increase in the share of block capacity in IPS 1 rises, but remains lower than the average one for UPS (see column 2 of table in Appendix D.11).

IPS 1 is defined also by the fact that it receives power over the interconnection ties in normal calculated operating state (mode) at a level of 1.5 / 2.0% of the IPS installed capacity at the corresponding time level (column 4 of table in Appendix D.11).

All the indicated factors provide in combination a relatively high operation reliability of IPS 1 in UPS. Table in Appendix D.12 shows that the mathematical expectation of the capacity under emergency outage (in p.u.) is lower than the average one for the whole UPS, and the share of the mathematical expectation of power undersupply in IPS 1 in the total undersupply for UPS is much lower than the share of the installed capacity.

It should also be underlined the following operation peculiarities of IPS 1:

- Despite the fact that the direct transmission capacity of ties in IPS 1 with the remaining system are the same in both directions at all time levels, distinctions in development of the remaining network lead to some decrease in operation reliability of IPS 1 in the second variant at the first and second levels. At the third level, the operation reliability of IPS 1 in the second variant is somewhat higher than in variant 1;
- In passing from one calculated level to the other, the reliability indices slightly vary. Thereby, the reliability is lower for the second level than for the first but for the third level the reliability is higher as against the second level, but does not reach the first level in the first option and exceeds it in the second option (see figure in Appendix D.7).

Hence, it is possible to conclude that the operation reliability of IPS 1 in UPS is comparatively high.

IPS 2. The installed capacity of IPS 2 amounts to 21–22% of the UPS capacity at all calculated levels (see column 1 in table of Appendix D.11). The relative increases of capacity at each calculated level are close to the increases for UPS as a whole deviating by 1–2% from the latter (see column 3 in table of Appendix D.11). Thereby, the greater part of new capacity is block equipment with the corresponding reliability level.

Because of preferential commissioning of large-block equipment at the second and third levels, the relative component of block capacity increases at this period, although remains lower than the average one for the whole UPS (see column 3 of table in Appendix D.11). The growth of the block capacity share offers prerequisites for reduction of the average reliability level of equipment operation (see table in Appendix D.12).

IPS 2, just as IPS 1, receives power over the interconnection ties in the normal calculated state (mode) at a level of 4.5–5.5% of the installed capacity of the corresponding calculated level (see column 4 in table of Appendix D.11), but the scheme of ties of IPS 2 with the remaining system has a significantly complex structure and operation conditions. IPS 2 receives the capacity over some ties (with IPS 4 and IPS 5), and supplies it over the other ties. Therewith, the power received through tie between IPS 2 and IPS 4 is 5.5–7.5% from P^{g}_{install} IPS 2. The calculations show that the power transmitted over this tie in the normal calculated operating state (mode) is already nearly maximal and does not allow the use of available emergency reserve in IPS 4 and IPS 5 during failures in IPS 2.

The joint action of all noted factors taking into account of the chosen volume of generating capacity reserves, as is seen from the calculations, guarantees a comparatively high operation reliability of IPS 2 (see figure in Appendix D.8). For example, at the first and third levels, the share of the mathematical expectation of power undersupply in IPS 2 in the total undersupply in the whole UPS is lower than the share of installed capacity of IPS 2 and amounts to 10–15%. At the second level, the share of the mathematical expectation of undersupply in IPS 2 in the total undersupply in UPS increases to 23–32% primarily due to discrepancy in the network development level in both considered options and operation conditions of IPS 2. The values of reliability indices for IPS 2 are at the level of average values for UPS; however, in specific cases, they are lower. The following conclusion is that the reliability of IPS 2 at the second level can be improved by shifting the time of construction of individual ties to an earlier period. The additional calculations and the analysis are required to determine, in what direction and what transmission lines must be constructed earlier.

It is also necessary to pay attention to such factors as:

- Operation reliability of IPS 2 in the second variant of network development is lower than in the first option at all calculated levels;
- Reduction of the general level of equipment operation reliability because of the increasing share of block generating units in IPS 2 is compensated for by the growing capacity reserve and network development. Therefore, the operation reliability level of IPS 2 at the third level remains the same as at the first level, excluding a sharp drop at the second level.

IPS 3. This IPS is characterized by a large value of installed capacity comprising almost one-third (30–32%) of the UPS capacity at all calculated levels. As a result, the operation reliability of IPS 3 in parallel with such IPSs as IPS 2 and IPS 6 determines a reliability level of the whole UPS (above 70% of UPS generating capacity is installed in IPSs 2, 3 and 6). The relative increases of capacity increases at each calculated level are close to those for UPS deviating from the latter by ±(2–3)% (see column 3 in table of Appendix D.11). However, the increase in the block capacity with respect to the total capacity commissioned in IPS 3 at some calculated levels exceeds the corresponding indices for UPS. At the third level, almost all newly installed capacity in IPS 3 is block with the units of 300 MW and higher. This case decreases the overall reliability level of equipment operation, particularly, if the running-in factor is taken into consideration (see table in Appendix D.12), making it considerably lower than the average one for UPS and in comparison with other IPSs.

In the normal calculated operating state IPS 3 receives 1–2% of P^{g}_{install} on the ties between IPS.

The results of reliability calculations showed that:

- The considered factors of operation of IPS 3 in UPS are such that lead to the lowest reliability indices in comparison with other IPSs (see figures in Appendices D.4–D.13). Hence, the share of the mathematical expectation of power undersupply in IPS 3 in the total undersupply in the whole UPS is from 38 to 98% for different variants, calculated levels and forecasts of equipment operation reliability;
- Operation reliability of IPS 3 in the second variant is higher than in the first. This factor and the increasing reliability level in the second option of IPS 6 basically support higher reliable operation of the whole UPS in the second variant despite the decreasing reliability level of IPS 1 and IPS 2 in this case;
- From the reliability viewpoint, IPS 3 develops highly unevenly with the general trend toward decrease in operation reliability at the last level as compared to the first level, although at the second level the reliability is somewhat higher than at the first (see figures in Appendices D.7–D.13), which is typical of some other IPSs. Hence it follows that the suggested options should appropriately be adjusted to substantially improve an operation reliability level for IPS 3 in UPS.

IPS 4. The installed capacity of IPS 4 amounts to about 10% of the UPS capacity (see table of Appendix D.11). The capacity increases basically by installation of large-block units. On the whole, from the reliability viewpoint, the indices characterizing operation conditions of IPS 4 are not worse and sometimes much better than the average ones for UPS (see table in Appendix D.12).

In normal calculated states (modes), IPS 4 is a redundant system transmitting power to other IPSs at all levels (see the schemes of figures in Appendices D.2 and D.3). The calculations show that during emergencies in IPSs it is possible to utilize available reserves of other IPSs effectively and almost fully. This is sufficient to secure high operation reliability of IPS 4 in UPS in all researched conditions (see figures in Appendices D.4–D.13). On the other hand, due to the small generating capacity $\left(P^{g}_{\text{install}}\right)$ this IPS cannot essentially influence the UPS reliability.

IPS 5. From the reliability viewpoint, the operation conditions of this IPS are similar to those of IPS 4. The installed capacity of IPS 5 is 3–4% of the UPS capacity. All new capacity is commissioned by the block units of 300 MW and higher. As a result, the share of installed block capacity increases to 78% at the last considered level, which is much higher than the average level for UPS (see table in Appendix D.11). This fact leads to the high values of the mathematical expectation of capacity under emergency outage (see table D.12). In the normal calculated states (modes), IPS 5 is a redundant system at all time levels. Therefore, in the case of emergency outages of its capacity it is possible to utilize available reserves of other IPSs, which is seen from the calculations. As a result, the high operation reliability of IPS 5 in UPS is maintained in any specified variant of network development (see figures in Appendices D.4–D.12).

IPS 6. The installed capacity is 17–18% of the installed capacity of UPS at all calculated levels. New capacities are commissioned mainly by block units. Hence, the share of block capacity in P^{g}_{install} of IPS 6 is higher than on the average in UPS, and the mathematical expectation of capacity under emergency outage is higher than that of the analogous indices of some other IPSs (see table in Appendix D.12).

For IPS 6, it is also characteristic that it receives capacity by the ties between the IPS in the normal calculated states (modes) for all calculated levels. The capacity flow over the ties makes up 4–5% of the installed capacity of IPS 6.

The calculations show that the combined action of the indicated factors leads to the following:

- The operation reliability of IPS 6 tends toward sharp decrease at transition from one calculated level to another (see figure in Appendix D.12);
- Because of the low reliability of operation of IPS 6 leads to the fact that the mathematical expectation of undersupply in it is 20–45% of the total for the UPS of undersupply at the first and second levels, thereby reducing the overall reliability indicator for the UPS as a whole;
- The second variant of network development secures a slightly lower operation reliability of IPS 6 at the first level and a higher reliability at the second and third levels.

Thus, the presented tentative engineering assessment of the design researches of UPS development shows that the comparative analysis of some quantitative indices of IPS operation conditions, including the reliability indices, enables to reveal the "bottlenecks" of the variants and select rational ways for substantial improvement of power supply reliability of designed systems.

References

1. Aleksandrov IA, Kovalev GF, Rudenko YN (1973) Comparative analysis of reliability of various options for the development of electric power systems. In: Reports at the third all-union scientific and technical meeting on the stability and reliability of the energy systems of the USSR. Energia, Leningrad, pp 394–405 (in Russia)
2. Methodical instructions for designing the development of power systems (2003). Izd-vo NTs ENAS, Moscow (in Russia)
3. Rokotyan SS, Shapiro IM (eds) (1985) Handbook on the design of electric power systems. Energoatomizdat, Moscow, 352 p (in Russia)
4. Kudryashov GR, Rudenko YN, Fedotova GA (1970) Reliability indicators of boiler and turbine units of the electric power systems of Siberia. In: Reliability and profitability of power systems. Reports of the scientific and technical conference on summarizing the experience of design, construction and operation of power transmission lines and substations in the regions of Siberia and the Far East, vol 1. Science, Siberia Otd, Novosibirsk, pp 181–198 (in Russia)
5. Gladyshev GP, Gorin VP, Yakovlev GG (1975) Reliability of blocks of thermal power stations of the USA in 1960–1071. Energy Manag. Abroad. 2:pp 1–11 (in Russia)
6. Melnikov NA (1969) Electrical networks and systems. Energia, Moscow, 496 pp (in Russia)
7. Rokotyan SS (1963) Operational characteristics of ultrahigh voltage lines. Energy Econ. Abroad. 6:pp 23–27 (in Russia)

[illegible]

References

[illegible]

Conclusion

The researches described in the work can be divided into three basic stages:

1. Development and formulation of the key provisions of the approach to EPS reliability analysis.
2. Development of the methods and algorithms for assessment of reliability indices of complex power systems. Construction of the model and the corresponding software.
3. Performance of mass calculations of reliability for different development variants of real and idealized power systems and hereon, adjustment of the approach and formulation of practical recommendations on the degree, forms, and accuracy of accounting diverse factors of EPS operation and development in the analysis of power supply reliability.

The practical conclusion of the first stage was the publication of recommendations [1].

The developed methodic of comparative analysis of the reliability of variants for the development of EPS at the level of the UPS, and IPS can become one of the means of taking account of the reliability factor in the design of systems. This complements the experience and intuition of specialists by quantitative relationships. The identification of "bottlenecks" and the most rational way of adjusting options in order to improve reliability provide on the basis of these quantitative assessments. Thus, the suggested method for reliability analysis can contribute to more justified formation of promising EPS structures.

The second stage of scientific and practical researches logically emerges from the first stage, since the developed approach to reliability accounting at power system designing assumes availability of the proper software for researchers.

As for the methods for reliability assessment of complex power systems, the results of researches carried out made it possible to overcome completely or partially the difficulties arising in such assessment. In this regard, we propose:

1. The methods for reduction of the amount of calculated system states (the method of "screening" the deficit-free states; the methods for limitation of the amount of

G. F. Kovalev and L. M. Lebedeva, *Reliability of Power Systems*,
Power Systems, https://doi.org/10.1007/978-3-030-18736-1

members in series of distribution of equipment states; the criterion of calculation termination);
2. The "energy reliability characteristics" (ERChs) of ties which enable, first, to arrange a stepwise calculation of multi-nodal schemes; second to ensure the linkage of reliability calculations of various hierarchical levels of EPS; and third, to obtain information for the technical and economic analysis of ties themselves;
3. The principle of accounting for the operational strategy ("dispatch policy") of consumers' restrictions while minimizing the capacity deficit in postemergency regimes in the process of assessing the reliability of EPS and some other provisions mentioned in the text of this work.

An attempt to take into account all the most significant factors of the development and functioning of EPS is done in the proposed model of EPS designed to assess reliability. The assumptions made in this case even if influence the results, however, considerably facilitate the solution of the set of very complex problem.

The main distinctive features of the constructed model are:

1. Limitation on the power system network structure is eliminated. The network can be of any configuration with the ties of limited transfer capability and the probabilities of transmission line failures.
2. The distribution series of generating capacity states can be calculated without preliminary equivalenting of the composition of units. This fact reduces labor inputs for preparation of initial information and improves calculation accuracy, particularly for power systems with great number units of different types.
3. The model can specify states of individual types of generating equipment and states of ties in the initial information by the corresponding distribution series.
4. Minimization of capacity deficits, taking into account the operational "policy" of limiting consumers, allows obtaining reliable indicators of reliability of electricity supply for both the system as a whole and its individual energy areas.
5. The reliability indices can be assessed for the whole calculated period and its individual intervals.

The listed features of the developed model allow, in addition to the main factors, to take into account many "nuances" of the development and functioning of EPS:

- process of running-in and aging of equipment (through the appropriate assignment of statistical values of q for certain accounting periods);
- special schemes of equipment operation (double blocks, special schemes for linking power lines and the influence of internal ties of the IPS on the interconnection capacity);
- seasonal and other non-stationarity of the equipment composition and characteristics of the system elements, including reliability characteristics;
- accounting for the time shifts in time for various nodes of the EPS and others.

The development of the calculation approach and the corresponding computational algorithms was based on application of the most effective methods devised for power system reliability calculations, on the one hand, and modification of some known algorithms to improve their accuracy and reduce the time of their work. The calculation algorithms of distribution series of generating capacity, ties, and others can serve as an example (see Chap. 3).

The designed programs are characterized by the following features:

1. The initial information can be assigned in various forms and with a varying degree of accuracy. A wide scope of reliability indices, including the computation of complex indicators, is of great importance for the research of the reliability properties of EPS. They characterize the properties and operating conditions of EPS in a generalized.
2. The programs are written in the FORTRAN language that is most commonly used for energy calculations.
3. Modular construction of the program allows it to be easily changed during the process of improvement or modification.

The following conclusions can be drawn with respect to the third level:

1. The performed researches confirmed and in individual cases adjusted significance of some factors for assessment of power supply reliability.
2. The rational choice of the form and accuracy of initial data setting reduces requirements for the initial information preparation and decreases time of power system reliability calculations.
3. The reliability calculations and analysis of power system development options show that the suggested approach enables to conduct a sufficiently effective assessment of design works from the reliability viewpoint and outline ways of their correction on the basis of the obtained quantitative characteristics.

The performed research and the obtained positive results can serve as a source for further researches in this sphere: collection of the calculated information and then a deeper and extensive analysis of indices and factors influencing it on this basis. Needless to say, the suggested approach will also be improved with accumulation of the experience.

In the future, it is possible to study the relation of this approach with the economic methods for the analysis of power system development options and the key design tasks. Besides, the planned works will apparently contribute to development and formulation of the standard indices of power system operation reliability, which is a topical task of the current stage of power industry development.

[illegible]

[illegible]

[illegible]

[illegible]

[illegible]

[illegible]

[illegible]

[illegible]

[illegible]

[illegible]

[illegible]

Appendix A

A.1 Structure and Algorithms of the Program for Determination of Power System Reliability

A.1.1 Initial Data[1]

The calculation schemes are constructed on the basis of the data on the researched system: information on generating equipment and loads of the energy nodes of scheme, configuration, and parameters of network. The scheme consists of M calculation nodes, whose power supply reliability characteristics should be determined, and N calculated ties between them.

Here, the calculated tie n between nodes m' and m'' is understood as a combination of all transmission lines of each n-th tie $\left(n = \overline{1, N}\right)$, which are specified by their reliability and energy parameters (see below).

Generating equipment. The permanent set and parameters of power system equipment are specified for each energy node and each calculated interval, which are enumerated in the calendar sequence.

At the first calculated interval ($s = 1$), the equipment set is specified by the year beginning. Then, the equipment composition by the year beginning can be changed (the accounting of dismantling, reconstruction, commissioning of new units, etc.) for each s-th calculated interval ($s \geq 2$).

The standards for major and average[2] repairs $\bar{t}_{\text{major}\,im}$ are represented by the average annual outages time of the units in repair (in days) for each group. The standards on current repairs are represented in per cent of the rated unit capacity for each group.

The standards on major repairs are composition only for the equipment by the year beginning. Correspondingly, the schedule of major repair volumes for a year is

---

[1]From the authors' standpoint, some information will inevitably be repeated to characterize initial information.

[2]A combination of major and medium repairs is called further for short "major repairs".

G. F. Kovalev and L. M. Lebedeva, *Reliability of Power Systems*,
Power Systems, https://doi.org/10.1007/978-3-030-18736-1

worked out only for this equipment. However, for the dismantled and newly commissioned units of this year the major repairs are obviously not planned.

The failure rate of units is represented by the probability (the probable relative duration) of emergency outage (unavailability): $q_{i_{ms}} = \tau_{i_{ms}}/T_s$.

Load schedules. The loads are represented by setting A typical daily schedules (p.u.). Each a-th schedule $\left(\alpha = \overline{1, A}\right)$ is written in the astronomical time starting with the first night hour, and for nodes, it is used based on the zone time.

The data indicating the load schedules, which take place (A_{ms}), and the number of operation days v_{ams} during the s-th calculated interval for each (A_{ms}) of the schedules $\left(\alpha_{ms} = \overline{1, A_{ms}}\right)$ correspond to the m-th node for each s-th interval. The number of days of node operation with the same own schedules should be equal. If one daily schedules of one node with the number of days of operation by this schedules, v_{ams} conforms to several different schedules at other nodes with the number of days

$$v'_{ams} + v''_{ams} + \cdots = v_{ams},$$

and then, several "different" daily duration schedules $v'_{ams}, v''_{ams}, \ldots$ should be set for all nodes.

In accordance with the existing practice of power system designing, the load schedules are assigned in the following form:

- the power of regular (average) annual peak load for each energy node $P^{\mathrm{l}}_{\max m}$ in MW;
- the capacities of peak loads for different intervals $P^{\mathrm{l}}_{\mathrm{month\,max}\,ms}$ in the fractions ρ_{ms} of $P^{\mathrm{l}}_{\max m}$ taken equal to 1.0;
- the capacities of daily peak loads $P^{\mathrm{l}}_{\mathrm{daily\,max}\,msa}$ in the fractions ρ_{msa} of the power of peak load at the calculated interval $P^{\mathrm{l}}_{\mathrm{month\,max}\,msn}$ taken equal to 1.0;
- the hourly capacities of maximum (peak) loads of the daily schedules P^{l}_{msat} in the fractions ρ_{msat} of the capacity of daily peak load $P^{\mathrm{l}}_{\mathrm{daily\,max}\,msa}$ taken equal to 1.0 (see the typical daily schedules A).

The annual schedules of monthly peak loads for the nodes necessary to "fit" major repairs are formed from the set values of the load peaks of the calculated intervals and the number of months in each of them. It is assumed that the peak loads of all months entering the corresponding s-th interval are equal to $P^{\mathrm{l}}_{\mathrm{month\,max}\,ms}$.

In general, to simulate irregular oscillations, the number of terms of the distribution series, which is the same for all nodes $B_m = 7$, is given by arrays of corresponding probabilities and σ^{l}_m values for each of the nodes. As an example of a distribution function representing a stepwise approximation of a normal law with a standard deviation $\sigma^{\mathrm{l}} = 1\,\%$ [2], the following series with a number of steps of seven:

$P_{irreg\,\beta_m}$	1.03	1.02	1.01	1.00	0.99	0.98	0.97
q_{β_m}	0.006	0.0606	0.2417	0.3834	0.2417	0.0606	0.006

The probabilities of q_{β_m} of the corresponding relative load values $P'_{irreg\,\beta_m}$ can be calculated, for example, by the technique described in [3].

Transmission lines. The composition of the transmission lines is set separately for each tie in the calculated intervals. Preliminary, the composition of the power lines is given by the beginning of the year $(s = 1)$. For each calculation interval $(s \geq 2)$, changes can made with respect to the composition and parameters of the transmission lines given at the beginning of the year.

The information about transmission lines is assigned with the indication of its reference to this or that tie "n" $\left(i_{ns} = \overline{1, I_{ns}}\right)$.

The problem of capacity deficit minimization in terms of physical and technical properties of power systems is solved by setting the maximum admissible transfer capabilities $\overline{P}_{i_{ns}}$ and $\underline{P}_{i_{ns}}$ of transmission lines in forward and inverse directions of the calculated ties, in addition to H_{i_n} and $q_{0\,i_{ns}}$.

As is noted in Chap. 3, the groups of transmission lines can be specified by the corresponding distribution series of transmission line states.

Auxiliary information. The following auxiliary information should be set for the operation of the program:

(1) k_{valley}—utilization factor of the valley in the annual schedule of monthly peak loads for major repairs;
(2) the calculated "step" (the stage of capacity variation) of the distribution series of generating equipment state for each node—$P_{step\ m}$ (see the recommendations of Sect. 4.3.5);
(3) the calculation accuracy of distribution series of generating capacity—ε_q (about the essence of this value see description of the subblock "Calculation of the distribution series of generating capacity state of an individual node");
(4) the calculation accuracy of the balance equations ε_r and the auxiliary dual minimized functional ε_ψ in the block of capacity deficit minimization;
(5) information on the composition of ties for which the energy reliability characteristics (ERChs) are calculated;
(6) the calculated capacity step for energy reliability characteristics of ties $P_{step\,n}$;
(7) the calculated capacity steps to calculate distribution series of capacity deficit at each of M energy nodes $P_{def\ step\ m}$, $m = \overline{1, M}$ and in the whole system $P_{def\ step\ syst}$.

The algorithms of the calculation program for complex power system reliability indices are described below in the order corresponding to the location of subblocks in the program block scheme presented in Fig. 3.1 of the text.

A.1.2 *Control of Initial Information Input and Its Intermediate Processing*

The initial information is inserted in accordance with the instructions for user and the requirements of the FORTRAN language.

When reading some initial information in the program, control is performed, for example:

- The relative durations of the outage q should be lower than unity.
- The standards of the scheduled current repairs $\bar{\alpha}_{\text{curr rep } i_{ms}}$ should not exceed unity.
- The amount of energy nodes, ties, groups of equipment, units and transmission lines in the groups should be strictly integer numbers and so on.

The initial data is subjected to the following processing.

1. Conversion of load values from relative units to named ones is done using the following formulas:

$$\left.\begin{aligned} P^{\text{l}}_{\text{month max } ms} &= \rho_{ms} \cdot P^{\text{l}}_{\text{max } ms} \\ P^{\text{l}}_{\text{daily max } msa} &= \rho_{msa} \cdot P^{\text{l}}_{\text{month max } ms} \\ P^{\text{l}}_{msat} &= \rho_{msat} \cdot P^{\text{l}}_{\text{daily max } msa} \end{aligned}\right\} \tag{A.1}$$

2. The specified information is applied to determine the values needed for calculation of reliability indices or complex characteristics of the factors influencing reliability:

(a) The required power generation by node is determined by the hourly load schedules

$$W_m = \sum_{s=1}^{S} \sum_{a_{ms}=1}^{A_{ms}} \sum_{t=1}^{24} P^{\text{l}}_{msat} \cdot v_{a_{ms}} \text{ MW h}.$$

(b) The required power generation in the whole system

$$W_{\text{syst}} = \sum_{m=1}^{M} W_m.$$

(c) The joint absolute peak load of the system is determined from the expression

$$P^{\text{l}}_{\text{joint max}} = \max\left(\sum_{m=1}^{M} P^{\text{l}}_{msat}\right)$$

for $t = \overline{1, 24}$; $a_{ms} = \overline{1, A_{ms}}$; $s = \overline{1, S}$.

(d) The number of hours of use the absolute peak of load by nodes is determined by the formula

$$T^{\mathrm{l}}_{\max m} = W_m / P^{\mathrm{l}}_{\max m}.$$

(e) The number of hours of use of the joint peak of the system load is determined by the formula

$$T^{\mathrm{l}}_{\max \mathrm{syst}} = W_{\mathrm{syst}} / P^{\mathrm{l}}_{\mathrm{joint}\max}.$$

3. All the initial data and the data calculated by the program are presented in the tabular form.

A.1.3 Calculation of Complex Characteristics of the Factors Determining Reliability of Nodes and the Whole System

The complex indices determining configuration and parameters of load schedules include:

- the absolute peak loads of the nodes $P^{\mathrm{l}}_{\max m}$ and the whole system $P^{\mathrm{l}}_{\mathrm{joint}\max}$;
- the required power generation by the energy nodes W_m and the whole system W_{syst};
- the peak load utilization time $T^{\mathrm{l}}_{\max m}$ and $T^{\mathrm{l}}_{\max \mathrm{syst}}$.

The calculation of the indicated indices is presented above (A.1.2). If necessary, they can be used to determine other complex indices, which are derivatives of the data, namely:

- the average load value $P^{\mathrm{l}}_{\mathrm{average}\, m} = W_m / 8760$;
- the load factors, etc.

The impact of equipment emergency rate on reliability of nodes and the whole system is assessed using the expectation of the relative emergency outage $\bar{q}_{\mathrm{emerg}}$ of equipment as a complex index in accordance with formula (4.1) of the text.

A.1.4 Calculation of Conditional Hourly Loads of Energy Nodes

This problem can be solved on the basis of the following initial data for each node:

(1) the set of equipment as groups of units by the year beginning and changes in the equipment set with respect to the year beginning for the intervals s;

(2) the amount of units, rated capacities of units and standards on the current and major repairs for each group;
(3) the load schedules for typical days during a year and the data indicating the load schedules for the calculated interval, and also the number of days of operation during each interval according to a particular load schedule;
(4) the annual peak load schedules of the calculated intervals and the number of months in each interval.

The load values assigned in p.u. are converted preliminarily into the named units (MW) by the formula in (A.1.1).

The volumes of major repairs are determined in the following sequence:

– The "area" in MW day that is required for major repairs is calculated by the formula:

$$V_{\text{major}\,m} = \left(\sum_{jm=1}^{Jm} P^{\text{g}}_{jm} \cdot I_{jm} \cdot \bar{t}_{\text{major}_{jm}} \right) / K_{\text{outage}} \text{ MW day;} \tag{A.2}$$

– The schedule of monthly peaks of loading from the minimum value to the largest (absolute peak) is ordered (ranked). The ranking suggests preservation of information about the calendar month, to which one or another peak load belongs;
– The volumes of major repairs are determined for each month. For this purpose, the cycle is arranged to fill a valley in the ranked monthly peak load schedule (see Fig. A.1) by the formula:

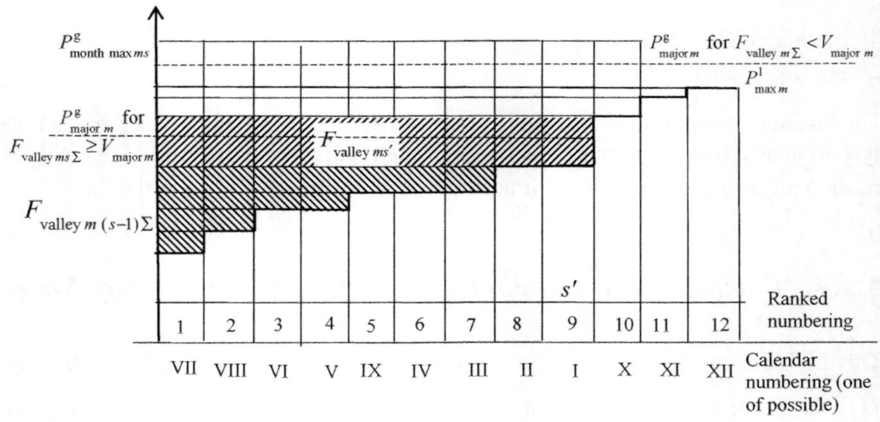

Fig. A.1 Ordered schedule of monthly peak loads and sequence of filling valley with major repairs

$$F_{\text{valley } ms'} = \left(P^{\text{l}}_{\text{month max } m(s'+1)} - P^{\text{l}}_{\text{month max } ms'}\right) \cdot 30.5 \cdot s'; \quad s' = \overline{1, 11}, \; \left(F_{\text{valley } m12} = 0\right) \tag{A.3}$$

The accumulated sums of separate monthly valleys are calculated at each step

$$F_{\text{valley } ms' \sum} = F_{\text{valley } m(s'-1) \sum} + F_{\text{valley } ms'} \tag{A.4}$$

and are compared to $V_{\text{major } m}$.

If $V_{\text{valley } ms' \sum} < V_{\text{major } m}$, then the cyclical process is continued; otherwise, the value $P^{\text{g}}_{\text{major } m} = \left(V_{\text{valley } m} - F_{\text{valley } m\,(s'-1) \sum}\right) / 30.5\, s'$ is determined and the capacity that is output to the major repair in each month from the 1st to the s' (referring to the ranking numbers, rather than the calendar sequence) is determined by the formula

$$P^{\text{g}}_{\text{major } m\chi} = \left(P^{\text{l}}_{\text{month max } ms'} - P^{\text{l}}_{\text{month max } m\chi}\right) + P^{\text{g}}_{\text{major } m}; \quad \chi = \overline{1, s'}. \tag{A.5}$$

In the rest of months, the major repairs are not planned

$$P^{\text{g}}_{\text{major } m\chi} = 0 \quad \text{for} \quad \chi > s'.$$

If it will be found that $V_{\text{valley } m \sum} < V_{\text{valley } m}$, where $F_{\text{valley } m \sum} = F_{\text{valley } m12 \sum}$, then $P^{\text{g}}_{\text{major } m}$ is determined by the expression

$$P^{\text{g}}_{\text{valley } m} = \left(V_{\text{major } m} - F_{\text{valley } m \sum}\right) / 365,$$

and the capacity, that is output for major repair each month of the year, is calculated by formula (A.5). In this formula,

$$P^{\text{l}}_{\text{month max } ms'} = P^{\text{l}}_{\text{max } m}, \text{ and } \chi = \overline{1, 12}.$$

The values of the calculated capacities $P^{\text{g}}_{\text{major } m\chi}$ are taken as corresponding to the calculated intervals in calendar numbering (sequence):

$$P^{\text{g}}_{\text{major } ms} = P^{\text{g}}_{\text{major } m\chi}; \quad s = \overline{1, S}, \; \chi = \overline{1, 12}. \tag{A.6}$$

The determination of the capacity for current repairs consists in calculating the power for the s-th interval

$$P^{\text{g}}_{\text{curr } ms} = \sum_{j_{ms}}^{J_{ms}} \overline{\alpha}_{\text{curr } j_{ms}}, \quad m = \overline{1, M}, \; s = \overline{1, S}. \tag{A.7}$$

The calculated values of the capacities for major and current repairs are added to the hourly loads of typical daily schedules of the corresponding calculated intervals:

$$P^{\mathrm{l\,cond}}_{msat} = P^{\mathrm{l}}_{msat} + P^{\mathrm{g}}_{\mathrm{major}\,ms} + P^{\mathrm{g}}_{\mathrm{curr}\,ms}\ \mathrm{MW},$$

where $t = \overline{1,24}$; $s = \overline{1,S}$; $\alpha = a_{ms} = \overline{1,A_{ms}}$.

For the s-th interval, every t-th hour of the a_{ms}-th daily schedule takes place $v_{st} = v_{a_{ms}}$ times (the same for all nodes). The indicated schedules are transformed to the ordered sequences for the algorithmic reasons.

The hourly total power of conditional loads of all nodes is taken as the ordering criterion (considering the time zone):

$$P^{\mathrm{l\,cond}}_{\sum sat} = \sum_{m=1}^{M} P^{\mathrm{l\,cond}}_{msat}, \quad t = \overline{1,24};\ a = \overline{1,A}. \tag{A.8}$$

All hourly load schedules are arranged in the decreasing order of $P^{\mathrm{l\,cond}}_{\sum sat}$.

The hours with equal loads in all subsystems are represented by one calculated subinterval φ_s with the duration equal to the sum of durations of combined hours

$$t_{\varphi_s} = \sum_{t_s^*} v_{st}, t_{st}^* \in \left\{t / P^{\mathrm{l\,cond}}_{msat} = \mathrm{idem}, \quad m = \overline{1,M}\right\}.$$

The output data of the subblock comprise an array of the ordered sequence of the conditional total loads $P^{\mathrm{l\,cond}}_{\sum\varphi_s} = P^{\mathrm{l\,cond}}_{\sum sat_s^*}$, corresponding arrays of the hourly conditional loads of the energy nodes $P^{\mathrm{l\,cond}}_{m\varphi_s} = P^{\mathrm{l\,cond}}_{msat_s^*}$ and durations of these loads t_{φ_s}.

The complex characteristics of conditional load schedules $P^{\mathrm{l\,cond}}_{\max m}$, $P^{\mathrm{l\,cond}}_{\mathrm{comb}\max}$, W^{cond}_{m}, $W^{\mathrm{cond}}_{\mathrm{syst}}$ are additionally calculated by the formulas presented in A.1.2.

A.1.5 Calculation of the Distribution Series of Generating Capacity States for an Individual Node

The distribution series of generating capacity states of a node is calculated by the generating function of the general theorem on repetition of experiments [3]:

$$\prod_{j_{ms}=1}^{J_{ms}} \left(p_{j_{ms}} + q_{j_{ms}}\left[P^{\mathrm{g}}_{j_{ms}}\right]\right)^{I_{j_{ms}}}. \tag{A.9}$$

Formula (A.9) is true for the joint and independent elementary events and is applicable here on the assumption that the power supply to consumers by any generator does not depend on the state of other system elements. Thus, this implies a parallel (in terms of reliability) operation of units, which corresponds to the accepted notion of an energy node as a "concentrated" one.

Calculation of the distribution series consists in decomposition of the generating function (A.9) into summands $q_{\text{emerg}}\left[P_d^{\text{g}}\right]$ which represent probabilities that the load is covered in the case the presence of the generating capacity P_d^{g}:

$$\prod_{j_{ms}=1}^{J_{ms}} \left(p_{j_{ms}} + q_{j_{ms}}\left[P_{j_{ms}}^{\text{g}}\right]\right)^{I_{j_{ms}}} = \sum_{d=1}^{D_{ms}} q_d\left[p_d^{\text{g}}\right] = 1, \tag{A.10}$$

where $D_{ms} = \sum_{i=0}^{I_{ms}} C_{I_{ms}}^{i} = 2^{I_{ms}}$.

In this formulation, the problem of determining the terms $q_d\left[P_d^{\text{g}}\right]$ for a large number of elements I_{ms} forming EPS is unsolvable because of its dimensionality, but the technical features of the object allow solving it in a way, proposed below.

The distribution series can be calculated for discrete values, each combining different equipment states on the principle of equal total capacity of normally operating generators. This is admissible for the "concentrated nodes," for which the generators "addresses" do not matter, but the total values of available capacities are of concern. In addition, the amount of calculated states of generating capacity substantially decreases.

To select the calculated power values of the series, an even uniform "step" (power change stage) of the series is set. The method for selecting the step is set out in Sect. 4.3.5. The number of terms of the series will be, in this case,

$$D'_{ms} = \text{entier}\left[P_{\text{install}\,ms}^{\text{g}}/P_{\text{step}\,m} + 1\right] \ll 2^{I_{ms}}.$$

As a rule, not all series members D'_{ms} are calculated, since the probability of low values of generating capacity, which is not in emergency outage, is extremely low at a great number of units. The number of distribution series members is limited on the ground of the required or possible calculation accuracy of the sought reliability indices using computer facilities.

As an example of limiting the number of calculated members of the distribution series, let us analyze the definition of one of the indicators of reliability—the relative supply of electricity to consumers $\pi = 1 - \text{M}W_{\text{und}}/W < 1$.

In particular, if the calculation accuracy of π is equal to nine digits after the point, the maximum admissible error of determining π will be $|\Delta\pi| = 10^{-10}$.

Hence, in this case it is inappropriate to calculate the values of $\text{M}W_{\text{und}}/W$ with an error lower than 10^{-10}, and MW_{und} with an error lower than $10^{-10} \times W$.

When some h-th member of the distribution series of generating capacity is achieved, the sum of these h members of series becomes so that the mathematical expectation of undersupply will be

$$\mathrm{MW}'_{\mathrm{und}} = \sum_{d=1}^{h} q_d\left[P_d^{\mathrm{g}}\right] \cdot \left(P_d^{\mathrm{l}} - P_d^{\mathrm{g}}\right) \cdot T_{\mathrm{p}} < 10^{-10} \times W.$$

With $P_d^{\mathrm{g}} \approx 0$, and the load at the d-th calculated state $P_d^{\mathrm{l}} \approx P_{\mathrm{max}}^{\mathrm{l}}$, then the condition for distribution series limitation (with a margin) will be represented as

$$\begin{aligned}\sum_{d=1}^{h} q_d[P_d^{\mathrm{g}}] \leq 10^{-10} \times W/\left(P_{\mathrm{max}}^{\mathrm{l}} \cdot T_{\mathrm{p}}\right) &= 10^{-10} \times \left(P_{\mathrm{average}}^{\mathrm{l}}/P_{\mathrm{max}}^{\mathrm{l}}\right) \\ &= k_{\mathrm{margin}}^{\mathrm{l}} \times 10^{-10},\end{aligned}$$

where $k_{\mathrm{margin}}^{\mathrm{l}} = P_{\mathrm{average}}^{\mathrm{l}}/P_{\mathrm{max}}^{\mathrm{l}}$ is factor "filling" the load schedule, the actual values of which are in the range from 0.5 to 1.0; therefore, the margin should be $\sum_{d=1}^{h} q_d\left[P_d^{\mathrm{g}}\right] \leq 10^{-11} < 0.5 \times 10^{-10}$.

In the general case, the condition for series limitation looks as follows

$$\sum_{d=1}^{h} q_d\left[P_d^{\mathrm{g}}\right] \leq 10^{-\varepsilon_q}. \tag{A.11}$$

Thus, to limit the number of members of the series, it is necessary to assign the appropriately chosen—ε_q.[3]

Because of divergence of some capacity values of units with the chosen calculated step $P_{\mathrm{step}\,m}$ of distribution series, it is necessary to solve the problem of placing the calculated probability value for any value of the total generating capacity P_d^{g} among the adjacent capacity stages of the sought series. This is realized by the known method of linear interpolation ("principle of moments") [4] in the following way.

A smaller contiguous number of a member of the sought $k = \mathrm{entier}\left[\frac{P_d^{\mathrm{g}}}{P_{\mathrm{step}\,m}}\right]$ series is determined for a capacity value P_d^{g} and its corresponding probability $q_d\left[P_d^{\mathrm{g}}\right]$.

The probabilities of the k-th and $(k+1)$—the members of the series add up from the shares of probabilities $q_d\left[P_d^{\mathrm{g}}\right]$, calculated by the formulas:

$$\left.\begin{aligned} q_k\left[P_k^{\mathrm{g}}\right] &= q_d\left[P_d^{\mathrm{g}}\right] \cdot \frac{P_{\mathrm{step}\,m} \cdot (k+1) - P_d^{\mathrm{g}}}{P_{\mathrm{step}\,m}} \\ q_{(k+1)}\left[P_{(k+1)}^{\mathrm{g}}\right] &= q_d\left[P_d^{\mathrm{g}}\right] \cdot \frac{P_d^{\mathrm{g}} - P_{\mathrm{step}\,m} \cdot k}{P_{\mathrm{step}\,m}} \end{aligned}\right\} \tag{A.12}$$

The initial data for calculation of the distribution series include the values of $P_{j_{ms}}^{\mathrm{g}}$, $I_{j_{ms}}$ and $q_{j_{ms}}$ for all J_{ms} groups of homogeneous units at the s-th interval.

The data for groups of units can be assigned in any of their sequence.

---

[3]From Sect. 4.3.4, it follows that ε_q is sufficient to take equal to 3–5.

The values of the calculated step $P_{\text{step}\,m}$ and the criterion for limitation of the series ε_q are assigned in addition. The number of series members D'_{ms} and the probabilities of failure-free equipment operation $p_{j_{ms}} = 1 - q_{j_{ms}}$ are determined in accordance with these conditions.

The resultant distribution series of generating capacity states at the node is calculated as follows.

In accordance with the formula of binomial distribution

$$Q_i^{\text{g}} = C_{I_{jms}}^i \cdot p_{j_{ms}}^{(I_{jms}-i)} \cdot q_{j_{ms}}^i; \quad i = \overline{0, I_{j_{ms}}}, \; p_{j_{ms}} = 1 - q_{j_{ms}},$$

the distribution series of states is calculated for the first group of units. The number of members of such a series is equal to $I_{j_{ms}} + 1$.

The first series member corresponding to the probability that all generating capacity is in operation (the number of failed units $i = 0$) is equal to $Q_0^{\text{g}} = p_{j_{ms}}^{I_{jms}}$.

All subsequent members are calculated by the recurrence formula.

$$Q_{i+1}^{\text{g}} = Q_i^{\text{g}} \frac{I_{j_{ms}} - i + 1}{i+1} \cdot \frac{q_{j_{ms}}}{p_{j_{ms}}}$$

In this case, the amount of capacity in the operable state—P_i^{g}—is equal to $P_i^{\text{g}} = P_{j_{ms}}^{\text{g}} \left(I_{j_{ms}} - i\right)$.

Each obtained value of Q_i^{g} is distributed between the cells k and $k + 1$ according to formulas (A.12).

As a result, the distribution series of generating capacity states is obtained for the units of the first group with the given step.

The distribution series of states of the units in the second group is calculated in a similar way. Then, by multiplying series with each other, we obtain a certain resultant series, which also has a given capacity step.

Further, the series of distributions of the next group of units is again calculated and multiplied with the obtained resultant series and so on until the last specified group of units.

The restriction of the number of members of the resulting series is carried out according to the expression (A.11). The summation of the discarded members of the series is carried out starting from the first. The sum obtained in such a way is added to the first remained series member (h).

When calculating the members of the series and performing the restriction procedure (A.11), the summation of numbers with significantly different orders (from 100 to 10^{-15}) is performed. This can lead to noticeable errors. To reduce the error, the summation is carried out with double precision.

Besides the distribution series, the mathematical expectation and the root-mean-square deviation of failure-free generating capacity state are calculated:

$$\mathrm{M}P'_{ms} = \sum_{j_{ms}=1}^{J_{ms}} P^{\mathrm{g}}_{j_{ms}} \cdot I_{j_{ms}} \cdot p_{j_{ms}}; \quad \sigma P'_{ms} = \sqrt{\sum_{j_{ms}=1}^{J_{ms}} \left(P^{\mathrm{g}}_{j_{ms}} - \mathrm{M}P'_{ms}\right)^2 \cdot I_{j_{ms}} \cdot q_{j_{ms}}}.$$

The output information of this subblock is:

- The values of distribution series members (probabilities) from h to K with the given capacity step in the sequence from the maximum equipment outages to the minimum ones. The number of series members is $K'_{ms} = K - h + 1$.
- The mathematical expectation and the root-mean-square deviation of the failure-free generating capacity state.

A.1.6 Multiplication of Distribution Series of Generating Capacity States at Nodes

Any two power distribution series given as a sequence of probabilities $\left\{q_{k'_{ms}}\left[P^{\mathrm{g}}_{k'_{ms}}\right]\right\}$ and $\left\{q_{k''_{ms}}\left[P^{\mathrm{g}}_{k''_{ms}}\right]\right\}$ with the corresponding power step $P'_{\mathrm{step}\,m}$ and $P''_{\mathrm{step}\,m}$ are the initial information for the multiplication subroutine of the distribution series. The calculated power step $P_{\mathrm{step}\,m}$ is set to calculate the resultant series. To reduce the number of members of the series at the level of K'_{ms} or K''_{ms}, $P_{\mathrm{step}\,m} = P'_{\mathrm{step}\,m} + P''_{\mathrm{step}\,m}$ is accepted.

The calculation process comprises:

- Determination of the number of resultant series members (if necessary)

$$K_{ms} = \mathrm{entier}\left[\left(P'_{\mathrm{step}\,m} \cdot K'_{ms} + P''_{\mathrm{step}\,m} \cdot K''_{ms}\right)/P_{\mathrm{step}\,m} + 1\right],$$

  where K'_{ms} and K''_{ms} are the numbers of the series members $\left\{q_{k'_{ms}}\left[P^{\mathrm{g}}_{k'_{ms}}\right]\right\}$ and $\left\{q_{k''_{ms}}\left[P^{\mathrm{g}}_{k''_{ms}}\right]\right\}$;
- Cyclical multiplication of each member of the first series by each member of the second series

$$q_{k_{ms}}\left[P^{\mathrm{g}}_{k_{ms}}\right] = q_{k'_{ms}}\left[P^{\mathrm{g}}_{k'_{ms}}\right] \cdot q_{k''_{ms}}\left[P^{\mathrm{g}}_{k''_{ms}}\right], \; k'_{ms} = \overline{1, K'_{ms}} \text{ and}$$
$$k''_{ms} = \overline{1, K''_{ms}}, \; P^{\mathrm{g}}_{k_{ms}} = P^{\mathrm{g}}_{k'_{ms}} + P^{\mathrm{g}}_{k''_{ms}};$$

the calculated values of probabilities for the capacities P_k^{g} are distributed among the stages of the capacity $P_{k_{ms}}^{\mathrm{g}}$, which correspond to the step $P_{\mathrm{step}\,m}^{\mathrm{g}}$ assumed for the resultant series. Distribution is carried out according to the rule of "moments" (A.12). The number of members of the series is not restricted.

– The total values of the mathematical expectation and the root-mean-square deviation of failure-free generating capacity state are calculated by the formulas:

$$\mathrm{M}P_{ms} = \mathrm{M}P'_{ms} + \mathrm{M}P''_{ms} \quad \left(\mathrm{M}P_{ms} = \sum_{k_{ms}=1}^{K_{ms}} P_{k_{ms}}^{\mathrm{g}} \cdot q_{k_{ms}}\left[P_{k_{ms}}^{\mathrm{g}}\right]\right);$$

$$\sigma P_{ms} = \sigma P'_{ms} + \sigma P''_{ms}; \quad \left(\sigma P_{ms} = \sqrt{\sum_{k_{ms}=1}^{K_{ms}} \left(P_{k_{ms}}^{\mathrm{g}} - \mathrm{M}P_{ms}\right)^2 \cdot q_{k_{ms}}\left[P_{k_{ms}}^{\mathrm{g}}\right]}\right).$$

A new distribution series can be multiplied by the next distribution series similarly and so on.

The output information includes:

- an array of the probabilities $\left\{q_{k_{ms}}\left[P_{k_{ms}}^{\mathrm{g}}\right]\right\}$ with the known calculated capacity step $P_{\mathrm{step}\,m}$;
- an amount of the series members K_{ms};
- an mathematical expectation and a root-mean-square deviation of the failure-free capacity state $\mathrm{M}P_{ms}$ and σP_{ms}.

A.1.7 Calculation of the Distribution Series of Transmission Lines States in a Separate Tie

In the specified set of transmission lines, each i_{ns}-th line ($i_{ns} = \overline{1, I_{ns}}$) is characterized by the length H_{i_n} and the probability accidents $q_{0i_{ns}}$.

The series are calculated separately for each tie "n" $n = \overline{1, N}$.

The values of outage probability

$$q_{i_{ns}} = q_{0i_{ns}} \cdot H_{i_n}/100$$

and the probabilities of its serviceable state $p_{i_{ns}} = 1 - q_{i_{ns}}$ are determined for each transmission line.

The distribution series of transmission line states is calculated using (A.10). In this case, each value $q_{k_{ns}}$ [q_d in the symbols of formula (A.10)] correlates with the specific state of lines $P_{k_{ns}}^{\mathrm{tl}}$ [P_d^{tl} in the symbols of formula (A.10)], which is formed by the procedure of "generator of states" available in the program of calculation the distribution series.

The values of $q_{k_{ns}}\left[P^{\text{tl}}_{k_{ns}}\right]$ are calculated by the formula

$$q_{k_{ns}}\left[P^{\text{tl}}_{k_{ns}}\right] = \left(\prod_{i_{ns}=1}^{I_{ns}} q_{i_{ns}}\right) \cdot \prod_{i_{ns}=1}^{I_{ns}} \eta_{i_{ns}k_{ns}},$$

where $k_{ns} = \overline{1, K_{ns}}$; $\eta_{i_{ns}k_{ns}} = 1$, if $\delta_{i_{ns}k_{ns}} = 0$;

$\eta_{i_{ns}k_{ns}} = p_{i_{ns}}/q_{i_{ns}}$, if $\delta_{i_{ns}k_{ns}} = 1$.

The parameter $\delta_{i_{ns}k_{ns}}$ is determined by the k_{ns}-th line states obtained with the help of the "generator of states":

$$\Delta^{\text{tl}}_{k_{ns}} = \{\delta_{1k_{ns}}, \delta_{2k_{ns}}, \ldots, \delta_{I_{ns}k_{ns}}\}.$$

Here, $\delta_{i_{ns}k_{ns}} = 1$, if the i_{ns}-th transmission line in the k_{ns}-th state is in a serviceable state, $\delta_{i_{ns}k_{ns}} = 0$, if the transmission line is in outage.

The transfer capabilities of ties (for use in the block of capacity deficit minimization—"CCD") are calculated by the formulas:

$$P^{\text{tl}}_{k_{ns}} = \{\underline{P}_{k_{ns}}, \overline{P}_{k_{ns}}\}; \quad \underline{P}_{k_{ns}} = \sum_{i_{ns}=1}^{I_{ns}} \underline{P}_{i_{ns}} \cdot \delta_{i_{ns}k_{ns}}; \quad \overline{P}_{k_{ns}} = \sum_{i_{ns}=1}^{I_{ns}} \overline{P}_{i_{ns}} \cdot \delta_{i_{ns}k_{ns}}.$$

The amount of calculated series members is limited by the assigned maximum amount of transmission lines I_{lim}, whose simultaneous outage in one tie is still sufficiently probable.

If $I_{ns} < I_{\text{lim}}$, then all members of the series are calculated by the formula

$$K_{ns} = \sum_{i=0}^{I_{ns}} C^{i}_{I_{ns}} = 2^{I_{ns}}.$$

If $I_{ns} > I_{\text{lim}}$, the amount of series members will be $K'_{ns} = \sum_{i=0}^{I_{\text{lim}}} C^{i}_{I_{ns}}$.

As a rule, I_{lim} is taken equal to 2–3 for one tie (see Sect. 4.3.6).

The output information of this subblock contains the probabilities $q_{k_{ns}}\left[P^{\text{tl}}_{k_{ns}}\right]$, the corresponding line states, and the number of these states that are characterized by the $\underline{P}_{k_{ns}}, \overline{P}_{k_{ns}}$ transfer capabilities. For the general algorithm of the program, it is required to generate line states in a sequence from the states of the maximum disabled number of transmission lines (minimum transfer capability) to the state of their full operability (maximum transfer capability).

A.1.8 Multiplication of Distribution Series of Transmission Line States

The initial information for the multiplication procedure of distribution series of transmission line states is any two distribution series specified as arrays of the probabilities $\left\{q_{k'_{ns}}\left[P^{\rm tl}_{k'_{ns}}\right]\right\}$ and $\left\{q_{k''_{ns}}\left[P^{\rm tl}_{k''_{ns}}\right]\right\}$ with the corresponding transfer capabilities $P^{\rm tl}_{k'_{ns}} = \left\{\underline{P}^{\rm tl}_{k'_{ns}}, \overline{P}^{\rm tl}_{k'_{ns}}\right\}$ and $P^{\rm tl}_{k''_{ns}} = \left\{\underline{P}^{\rm tl}_{k''_{ns}}, \overline{P}^{\rm tl}_{k''_{ns}}\right\}$.

The calculation process includes a cyclical multiplication of each member of the first series by each member of the second series

$$q_{k_{ns}}\left[P^{\rm tl}_{k_{ns}}\right] = q_{k'_{ns}}\left[P^{\rm tl}_{k'_{ns}}\right] \cdot q_{k''_{ns}}\left[P^{\rm tl}_{k''_{ns}}\right]; \quad k'_{ns} = \overline{1, K'_{ns}};\ k''_{ns} = \overline{1, K''_{ns}};$$
$$P^{\rm tl}_{k_{ns}} = P^{\rm tl}_{k'_{ns}} + P^{\rm tl}_{k''_{ns}};\ \left(\underline{P}^{\rm tl}_{k_{ns}} = \underline{P}^{\rm tl}_{k'_{ns}} + \underline{P}^{\rm tl}_{k''_{ns}};\ \overline{P}^{\rm tl}_{k_{ns}} = \overline{P}^{\rm tl}_{k'_{ns}} + \underline{P}^{\rm tl}_{k''_{ns}}\right);$$

If necessary, a new distribution series can be multiplied by the next series similarly and so forth.

The output information of this subblock comprises the probabilities $q_{k_{ns}}\left[P^{\rm tl}_{k_{ns}}\right]$, and the corresponding transfer capabilities $\underline{P}^{\rm tl}_{k_{ns}}$ and $\overline{P}^{\rm tl}_{k_{ns}}$ arranged in the sequence from the minimum to the maximum values. After summation of the probabilities, the amount of the resultant series members for the states with equal transfer capabilities will be $K_{ns} \leq K'_{ns} \cdot K''_{ns}$.

A.1.9 Block of "Screening" Deficit-Free System States and Choice of a New Calculated State

The principle of work of the described program algorithm is represented by the scheme in Fig. A.2.

In this scheme, the symbols φ_s, β and k_s denote:

The φ_s*-th* state of the calculated hourly loads of **nodes**, which comply with the load values of the array $P^{\rm l\,cond}_{\Sigma\varphi_s}(\varphi_s) = \overline{1, \Phi}$

$$P^{\rm l\,cond}_{\sum 1_s}\left(P^{\rm l\,cond}_{\rm comb\,max}\right), P^{\rm l\,cond}_{\sum 2_s}, \ldots, P^{\rm l\,cond}_{\sum \varphi_s}\left(P^{\rm l\,cond}_{\rm comb\,min}\right); \tag{A.13}$$

The β-th state of the irregular loads at energy nodes, $\beta = \overline{1, B}$, $B = \prod_{m=1}^{M} B_m = B_m^M$.

The irregular fluctuations are determined by the distribution series of the form:

$$q_1, q_2, \ldots, q_{\beta m}, \ldots, q_{Bm}; \quad \beta_m = \overline{1, B_m},\ m = \overline{1, M}$$

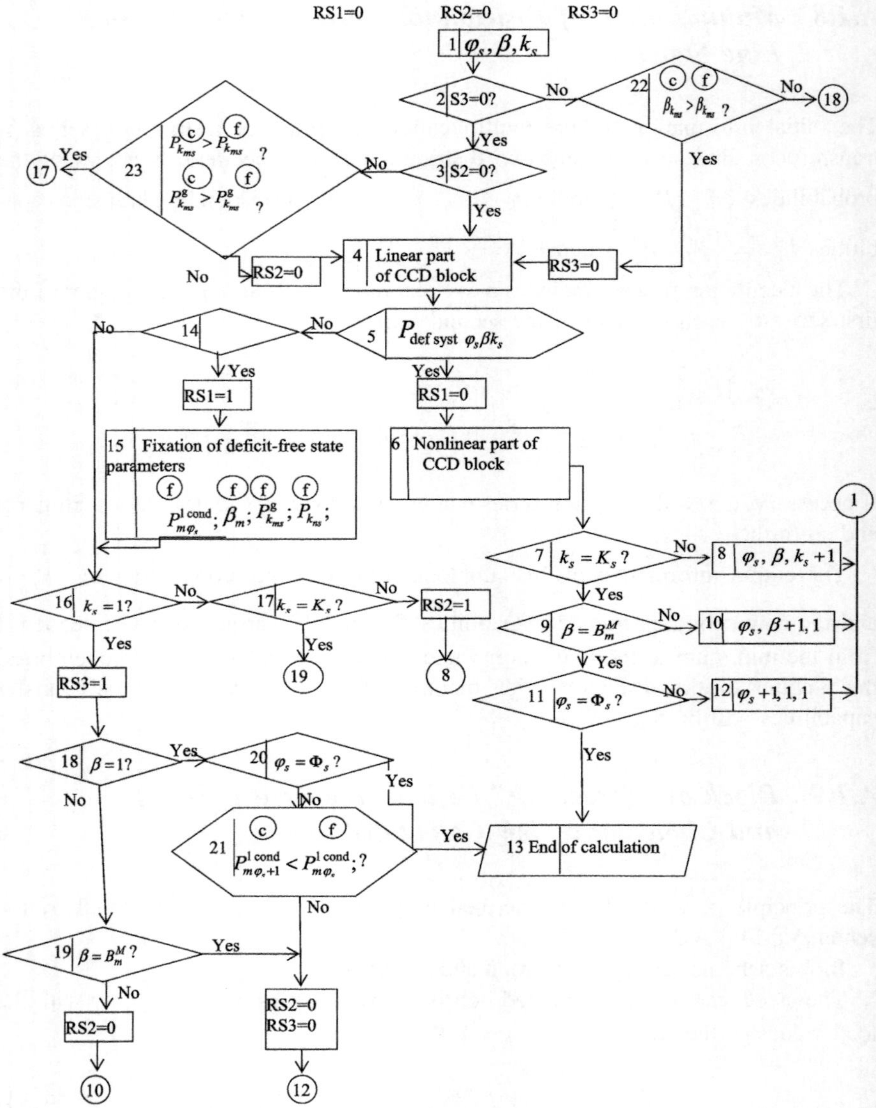

Fig. A.2 Logical scheme of "screening" deficit-free calculated states

All M series have the same amount of members (B_m) and the same probabilities for all energy nodes in accordance with the accepted method of their setting in the initial data. The values of irregular conditional loads are determined by the formula

$$P^{\mathrm{l\,cond}}_{m\varphi_s\beta} = P^{\mathrm{l\,cond}}_{m\varphi_s} \cdot \left(1 + \sigma^{\mathrm{l}}_m\left(\frac{B_m - 1}{2} + 1 - \beta_m\right)\right); \tag{A.14}$$

The k_s-th calculated state of system equipment, $k_s = \overline{1, K_s}$,

where $K_s = \prod_{m=1}^{M} K_{ms} \cdot \prod_{n=1}^{N} K_{ns} = \prod_{l=1}^{L} K_{ls}$.

The distribution series of generating capacity states by nodes and transmission lines states by ties have the form $q_{1_{l_s}}, q_{2_{l_s}}, q_{3_{l_s}}, \ldots, q_{k_{l_s}}, \ldots, q_{K_{l_s}}$.

In this case, $l = \overline{1, L}, L = M + N,\ l = m$ for the energy node $m = \overline{1, M}$ and $l = M + n$ for the ties $n = \overline{1, N}$.

The calculated states of system *equipment* and the combinations of irregular load values by node are determined by the cyclical enumeration of all possible combinations; the probabilities of these states are products of probabilities of the corresponding members of all distribution series.

In order to make each next state really lighter than the previous one, these states are enumerated in a special way: at first the heaviest one, which is determined by the first members of series, then the states, which are determined only by the first and by the second members of series, excluding the considered ones; then by the third members of series, and so on.

The purposeful enumeration of calculated system states is performed by two "state generators" called "simple" and "variable," which are constructed in the program. The "simple" state generator is intended for generation of states of irregular load fluctuations as combinations of these fluctuations by node from the heaviest state to the lightest one.

The states are generated using the rectangular matrix $M \times B_m$ of the assigned values of load fluctuations at each of M nodes; in this case, $B_m = \mathrm{const}, \quad m = \overline{1, M}$.

The "variable" generator of states creates equipment states for the whole system as combinations of all earlier determined states of generating capacities at the nodes and the states of transmission lines in the ties in the required sequence (as the "heaviness" decreases). The states are generated using the matrix L of the rows of different length[4]:

$$K_{l_s} = \mathrm{var}, \quad l = \overline{1, L}.$$

However, it should be noted that the formal data contained in the matrices $M \times B_m$ and $L \times K_{l_s}$ are not always sufficient for assessing the true state heaviness. The final

[4]The "variable" state generator might be applied for generation of the states of irregular fluctuations. However, its flexibility is time-consuming; therefore, it was thought expedient to insert the less complicated and faster "simple" state generator in the program.

conclusion is possible only on the basis of mode calculation. Therefore, in principle, decreasing sequences of severity can be violated by separate transitions to heavier states.

This fact is taken into consideration in the algorithm of "screening" deficit-free states by additional tests for shortage of the indicated states (see Fig. A.2).

The deficit-free states are "screened" by comparison of the next considered states with some fixed states (the next (current) states in the scheme of Fig. A.2 are indicated by index "c", and the fixed states—by index "f"). The fixed state corresponds to the last shortage-free system state determined before the considered stage of the cyclical enumeration of all states.

One solution or the other is accepted depending on the relationships among the initial parameters of the current and fixed states. If

$$\overset{\textcircled{c}}{P}{}^{1\,\mathrm{cond}}_{m\varphi_s} \le \overset{\textcircled{f}}{P}{}^{1\,\mathrm{cond}}_{m\varphi_s}; \quad \overset{\textcircled{c}}{\beta}_m \le \overset{\textcircled{f}}{\beta}_m; \quad \overset{\textcircled{c}}{P}{}^{\mathrm{g}}_{k\,ms} \le \overset{\textcircled{f}}{P}{}^{\mathrm{g}}_{k\,ms}; \quad m=\overline{1,M}; \quad \overset{\textcircled{c}}{\overline{P}}_{k\,ns} \le \overset{\textcircled{f}}{\overline{P}}_{k\,ns};$$

then it will evidently be a deficit-free state as well. In this case, the next seemingly deficit state is generated and tested.

Even if one of the conditions in (A.15) is not fulfilled, the given current state is calculated in the CCD block.

The logical variables *RS*1, *RS*2, and *RS*3 taking the value 0 or 1 are added to effectively check the relationships in (A.15). Before the block of system state generation starts to work, the zero values are assigned to the indicated variables.

After the first deficit-free state during calculation of current system states is obtained, it is "fixed" (its input parameters are stored), and *RS*1 is set equal to 1. If in the further work the system shortage state is obtained again, *RS*1 is set equal to 0. This value is retained, until the next shortage-free state is obtained, and so on.

Variables *RS*2 and *RS*3 serve for the organization of a shortened comparison of the current state with fixed not all the initial parameters (A.15), but only equipment only: if *RS*2 = 1, then:

$$\overset{\textcircled{c}}{P}{}^{\mathrm{g}}_{k_{ns}} \ge \overset{\textcircled{f}}{P}{}^{\mathrm{g}}_{k_{ns}}; \quad n=\overline{1,N}; \qquad \overset{\textcircled{c}}{P}{}^{\mathrm{g}}_{k_{ms}} \ge \overset{\textcircled{f}}{P}{}^{\mathrm{g}}_{k_{ms}}; \quad m=\overline{1,M};$$

or only for irregular load deviations: if *RS*3 = 1, then $\overset{\textcircled{c}}{\beta}_m \le \overset{\textcircled{f}}{\beta}_m;\ m=\overline{1,M}$ is checked, since other parameters will certainly contribute to the lack of deficiency of the current state in question.

The CCD block includes two subprograms of capacity deficit minimization by the linear and nonlinear programming methods. The calculation using the nonlinear programming method, which enables to distribute the total capacity deficit by nodes of the system according to the accepted criterion, is made after the calculation using the linear programming method, if as a result of this calculation the nonzero deficit in the system is obtained. Otherwise, this operating state is fixed as a deficit-free one that does not involve calculation using the nonlinear method in the CCD block, and the next state to be calculated by the program is chosen according to the above principles. The preliminary state estimation by the linear programming method decreases computing time of the block of system state generation. The amount of all possible system states for the s-th interval is determined as follows

$$G_s = \overline{\Phi}_s \cdot B_m^M \cdot \prod_{l=1}^{L} K_{l_s}.$$

This amount can obviously be very large, but as is mentioned above, the suggested logical scheme enables to partially "screening" deficit-free states and to switch from one calculated state to the other by the logical choice of the current values of the indices φ_s, β and k_s.

The value of the screening coefficient K_{scr} depends on the reliability level of the studied system. For the highly reliable system $\mathscr{P} > 0.999$, it is possible to decrease the number of calculated states by 3–4 orders of magnitude in comparison with $G_s(K_{\text{scr}} = 10^{-3}$–$10^{-4})$. For the low reliable systems, the screening effect will be weaker: $G_{s\,\text{fact}} = K_{\text{scr}} \cdot G_s$.

The probabilities of the corresponding states of loads and system equipment are calculated by formulas (2.12)–(2.14) of the main text.

The multiplied series members have, as a rule, negative exponents (up to 10^{-6}–10^{-8}). Therefore, the double accuracy allowing one to represent the numbers of a wide range is applied in the program in the FORTRAN language to determine probabilities of complex states.

The output information of the block comprises the values of t_{φ_s}, the probabilities of different irregular load values $q_{\varphi_{s\beta}}$, and equipment states q_{k_s} of each φ_s-th subinterval, as well as the corresponding values of generating capacities P^{g}_{kms} and load $P^{\text{l cond}}_{m\varphi_s\beta}$ for each m-th node and transfer capabilities $\underline{P}_{k_{ns}}$ and $\overline{P}_{k_{ns}}$ of all N ties.

A.1.10 Calculation of Energy Reliability Characteristic (ERCh) of Ties

The energy reliability characteristics (ERChs) of ties are calculated based on the information of the results of previous program subblocks: the subblock of system state generation and the subblock of capacity deficit minimization. This information includes:

- the probabilities of calculated states;
- flow distribution of capacity over ties in these states.

Additionally, the energy reliability characteristics can be obtained by setting the calculated step for capacity, for transfer capabilities. And also to set a list of ties whose energy reliability characteristics should be determined in the calculation scheme under consideration.

The principle of calculating the energy reliability characteristics consists in the following. The number of distribution series members of capacity flows in each tie, whose energy reliability characteristics are calculated, is determined on the basis of the specified calculated step $P_{\text{step}\,n}$ and the transfer capabilities $\underline{P}_n$ and $\overline{P}_n$:

$$K_n^{\text{ERCh}} = \text{entier}\left[(|\underline{P}_n| + |\bar{P}_n|)/P_{\text{step}\,n} + \alpha\right],$$

where $\alpha = 1, 2 \, \text{or} \, 3$ depending on the relationship among $\underline{P}_n, \overline{P}_n \, \text{and} \, P_{\text{step}\,n}$.

The probability values of corresponding power flows are distributed by the "principle of moments" in accordance with the capacity values of the series, which are determined by the calculated step.

The distribution series of power flows in the ties in physical conditions of power system operation described by the model are obtained after analysis of all calculated system states.

If all calculated system states (the sum of their probabilities is equal to 1) are analyzed, the overall energy reliability characteristic of the tie is obtained, which allows the assessment of performance effectiveness of this tie.

If only part of system states (shortage or near shortage) is analyzed, the sum of series members for each tie will be lower than unity and such characteristic can be applied only to further reliability calculations of the considered system provided that only these states will be analyzed subsequently.

For further application, the obtained characteristic, which represents unconditional probabilities of the onset of the considered states to occur in the system and the i-th capacity flow in the given tie—$\mathcal{P}(\text{uncond and } P_i)$—must be transformed to the characteristic representing conditional probabilities that in the considered system state, the flow in tie will be equal to P_i, i.e., to the probability $P_{\text{uncond}}(P_i)$. In this case, the unconditional probability of the analyzed states is

$$\mathcal{P}(\text{uncond}) = \sum_{r=1}^{R} \mathcal{P}(\text{uncond and } P_r)$$

where R is the amount of the considered system states.

The probability $P_{\text{uncond}}(P_i)$ is determined by the rule of multiplication of probabilities:

$$\mathcal{P}(\text{uncond and } P_r) = \mathcal{P}(\text{uncond}) \cdot P_{\text{uncond}}(P_i)$$

Hence,

$$P_{\text{uncond}}(P_i) = \mathscr{P}(\text{uncond and } P_r)/\mathscr{P}(\text{uncond})$$
$$= \mathscr{P}(\text{uncond and } P_r)/\sum_{r=1}^{R}\mathscr{P}(\text{uncond and } P_r)$$

The output information of this block is the arrays of the values of the probabilities of flows over the ties with the indication of the series step and the tie to which the corresponding energy reliability characteristic (ERCh) belongs.

A.1.11 Calculation of Reliability Indices for Nodes and the Whole System

The initial information for calculation of reliability indices of nodes in the system and reliability of the whole system comprises:

- the capacity deficit $P_{\text{def}\, m\, \varphi_s\beta k_s} = \Delta P^0_{m\, \varphi_s\beta k_s} > 0$ at the corresponding nodes in the whole system $P_{\text{def syst}\, \varphi_s\beta k_s} = \Delta^0_{\text{syst}\, \varphi_s\beta k_s}$ for the k_s-th states of system equipment and for the $\varphi_s\beta$-th loads of the s-th interval;
- the probabilities of shortage states of the nodes $q_{\text{def}\, m\varphi_s\beta k_s}$ and the system $q_{\text{def syst}\, \varphi_s\beta k_s}$, equal to the product of the probabilities $q_{\varphi_s\beta} \cdot q_{k_s}$ of the corresponding states of loads and equipment.

The specific feature of the block for calculation of reliability indices is that the values of $q_{\text{def}\, m\varphi_s\beta k_s}$ $\left(q_{\text{def syst}\, \varphi_s\beta k_s}\right)$ and $P_{\text{def}\, m\varphi_s\beta k_s}$ $\left(P_{\text{def syst}\, \varphi_s\beta k_s}\right)$ obtained after calculation of each deficit state are not stored, but immediately used for determination of the reliability indices.

The calculated indices can be divided into the indices of two levels: the "primary" indices determined directly based on the values of $q_{\text{def}\, m\varphi_s\beta k_s}$, $\left(q_{\text{def syst}\, \varphi_s\beta k_s}\right)$, $P_{\text{def}\, m\varphi_s\beta k_s}$, and $\left(P_{\text{def syst}\, \varphi_s\beta k_s}\right)$ and the "secondary" indices determined by the values of the primary indices and the values of the parameters of power system operation.

The primary indices include:

- the probabilities of the m-th node to be in the deficit states at the φ_s-th subinterval of the s-th interval

$$Q_{m\varphi_s} = \sum_{\beta=1}^{B}\sum_{k_s=1}^{K_s} q_{\text{def}\, m\varphi_s\beta k_s}, \quad m = \overline{1, M};\ \varphi_s = \overline{1, \Phi_s};\ s = \overline{1, S};$$

- the probabilities of the system to be in the shortage states at the φ_s-th subinterval of the s-th interval

$$Q_{\text{syst}\,\varphi_s} = \sum_{\beta=1}^{B} \sum_{k_s=1}^{K_s} q_{\text{def syst}\,\varphi_s \beta k_s}, \quad \varphi_s = \overline{1, \Phi_s};\ s = \overline{1, S};$$

- the distribution series of capacity deficit for each of M nodes, which are calculated by the values of $q_{\text{def}\,m\varphi_s\beta k_s}$ and $P_{\text{def}\,m\varphi_s\beta k_s}$ at the specified step of the $P_{\text{def step}\,m}$ using the "principle of moments" for each interval s

$$q_{\text{def}\,msd} = q_{\text{def}\,msd}(P_{\text{def}\,msd}),$$

in this case

$$q_{\text{def}\,msd} = q_{\text{def}\,m\varphi_s\beta k_s} \cdot t_{\varphi_s}/T_s, \quad d = \overline{1, D_{\text{def}\,ms}},$$

and for the whole year

$$q_{\text{def}\,md} = q_{\text{def}\,md}(P_{\text{def}\,md}),$$

in this case

$$q_{\text{def}\,md} = q_{\text{def}\,m\varphi_s\beta k_s} \cdot t_{\varphi_s}/8760, \quad d = \overline{1, D_{\text{def}\,m}};$$

- the probability of the m-th node to be in the shortage states at the s-th interval

$$Q_{ms} = \left(\sum_{\varphi_s=1}^{\Phi_s} \left(t_{\varphi_s} \sum_{\beta=1}^{B} \sum_{k_s=1}^{K_s} q_{\text{def}\,m\varphi_s\beta k_s} \right) \right) / T_s, \quad m = \overline{1, M};$$

- the probability of each of M nodes to be in the shortage states during the whole year

$$Q_m = \left(\sum_{s=1}^{S} \sum_{\varphi_s=1}^{\Phi_s} \left(t_{\varphi_s} \sum_{\beta=1}^{B} \sum_{k_s=1}^{K_s} q_{\text{def}\,m\varphi_s\beta k_s} \right) \right) / 8760$$
$$= \sum_{s=1}^{S} Q_{ms} \cdot T_s/8760, \quad m = \overline{1, M};$$

- the distribution series of the capacity deficit in the whole system that are calculated from the values of $q_{\text{def syst}\,\varphi_s \beta k_s}$ and $P_{\text{def syst}\,\varphi_s \beta k_s}$ at the specified step of the power $P_{\text{def step syst}}$ using the "principle of moments" for each interval s:

$$q_{\text{def syst}\,sd} = q_{\text{def syst}\,sd}\left(P_{\text{def syst}\,sd}\right),$$

in this case

$$q_{\text{def syst}\,sd} = q_{\text{def syst}\,\varphi_s \beta k_s} \cdot t_{\varphi_s}/T_s, \quad d = \overline{1, D_{\text{def syst}\,s}},$$

and for the whole year

$$q_{\text{def syst}\,d} = q_{\text{def syst}\,d}\left(P_{\text{def syst}\,d}\right),$$

in this case

$$q_{\text{def syst}\,d} = q_{\text{def syst}\,\varphi_s \beta k_s} \cdot t_{\varphi_s}/8760, \quad d = \overline{1, D_{\text{def syst}}};$$

- the probability of the system to be in the deficit states at the s-th interval

$$Q_{\text{syst}\,s} = \left(\sum_{\varphi_s=1}^{\Phi_s}\left(t_{\varphi_s}\sum_{\beta=1}^{B}\sum_{k_s=1}^{K_s} q_{\text{def syst}\,\varphi_s \beta k_s}\right)\right)/T_s;$$

- the probability of the system to be in the deficit states during the whole year

$$Q_{\text{syst}} = \left(\sum_{s=1}^{S}\sum_{\varphi_s=1}^{\Phi_s}\left(t_{\varphi_s}\sum_{\beta=1}^{B}\sum_{k_s=1}^{K_s} q_{\text{def syst}\,\varphi_s \beta k_s}\right)\right)/8760 = \sum_{s=1}^{S} Q_{\text{syst}\,s} \cdot T_s/8760;$$

- the mathematical expectation of capacity deficit at each m-th node at the s-th interval;

- the mathematical expectation of capacity deficit in the system at the s-th interval:

$$\mathrm{M}P_{\mathrm{def\,syst}\,s} = \left(\sum_{\varphi_s=1}^{\Phi_s}\left(t_{\varphi_s}\sum_{\beta=1}^{B}\sum_{k_s=1}^{K_s} q_{\mathrm{def\,syst}\varphi_s\beta k_s}\cdot P_{\mathrm{def\,syst}\,\varphi_s\beta k_s}\right)\right)/T_s;$$

- the mathematical expectation of capacity deficit at each m-th node during the whole year:

$$\mathrm{M}P_{\mathrm{def}\,m} = \left(\sum_{s=1}^{S}\sum_{\varphi_s=1}^{\Phi_s}\left(t_{\varphi_s}\sum_{\beta=1}^{B}\sum_{k_s=1}^{K_s} q_{\mathrm{def}\,m\varphi_s\beta k_s}\cdot P_{\mathrm{def}\,m\varphi_s\beta k_s}\right)\right)/8760$$
$$= \sum_{s=1}^{S}\mathrm{M}P_{\mathrm{def}\,ms}\cdot T_s/8760;$$

- the mathematical expectation of capacity deficit in the system during the whole year:

$$\mathrm{M}P_{\mathrm{def\,syst}} = \left(\sum_{s=1}^{S}\sum_{\varphi_s=1}^{\Phi_s}\left(t_{\varphi_s}\sum_{\beta=1}^{B}\sum_{k_s=1}^{K_s} q_{\mathrm{def\,syst}\,\varphi_s\beta k_s}\cdot P_{\mathrm{def\,syst}\,\varphi_s\beta k_s}\right)\right)/8760$$
$$= \sum_{s=1}^{S}\mathrm{M}P_{\mathrm{def\,syst}\,s}\cdot T_s/8760;$$

The <u>secondary</u> indices comprise:

- the coefficients of failure-free power supply at each m-th node and in the whole system at the s-th interval (the availability factors):

$$\mathscr{P}_{ms} = 1 - Q_{ms},\quad m=\overline{1,M};\quad \mathscr{P}_{\mathrm{syst}\,s} = 1 - Q_{\mathrm{syst}\,s}$$

- the coefficients of failure-free power supply at each m-th node and.in the whole systemи during the year (the availability factors):

$$\mathscr{P}_{m} = 1 - Q_{m},\quad m=\overline{1,M};\quad \mathscr{P}_{\mathrm{syst}} = 1 - Q_{\mathrm{syst}}$$

- the average values undersupply of electric power at each m-th node and in the whole system in the s-th interval:

$$W_{\mathrm{und}\,ms} = \mathrm{M}P_{\mathrm{def}\,ms}\cdot T_s,\quad m=\overline{1,M};\quad W_{\mathrm{und\,syst}\,s} = \mathrm{M}P_{\mathrm{def\,syst}\,s}\cdot T_s;$$

- the average values of undersupply of electric power at each m-th node and in the whole system during the year:

$$W_{\text{und}\,m} = 8760 \cdot \text{M}P_{\text{def}\,m}, \quad m = \overline{1, M}; \quad W_{\text{und syst}} = 8760 \cdot \text{M}P_{\text{def syst}};$$

- the relative energy index of reliability at each m-th node and in the whole system at the s-th interval:

$$\pi_{ms} = 1 - \text{M}W_{\text{und}\,ms}/W_{ms}, \quad m = \overline{1, M}, \; s = \overline{1, S};$$
$$\pi_{\text{syst}\,s} = 1 - \text{M}W_{\text{und syst}\,s}/W_{\text{syst}\,s};$$

- the relative energy index of reliability at each m-th node and in the whole system during the year:

$$\pi_m = 1 - \text{M}W_{\text{und}\,m}/W_m, \quad m = \overline{1, M}; \quad \pi_{\text{syst}} = 1 - \text{M}W_{\text{und syst}}/W_{\text{syst}}.$$

Besides the mathematical expectations of random values applied as reliability indices, the root-mean-square deviations of these random values are calculated

$$\sigma\mathcal{P}_{ms} = \left(\sum_{\phi_s = 1}^{\Phi_s} \left(Q_{ms} - Q_{m\phi_s}\right)^2 \cdot t_{\phi_s}/T_s \right)^{0.5};$$

$$\sigma\mathcal{P}_{\text{syst}\,s} = \left(\sum_{\phi_s = 1}^{\Phi_s} \left(Q_{\text{syst}\,s} - Q_{\text{syst}\,\phi_s}\right)^2 \cdot t_{\phi_s}/T_s \right)^{0.5};$$

$$\sigma\mathcal{P}_m = \left(\sum_{s=1}^{S} \left(\sum_{\phi_s = 1}^{\Phi_s} \left(Q_{ms} - Q_{m\phi_s}\right)^2 \cdot t_{\phi_s} \right) /8760 \right)^{0.5}$$
$$= \left(\left(\sum_{s=1}^{S} \sigma\mathcal{P}_{ms} \right)^2 \cdot T_s/8760 \right)^{0.5};$$

$$\sigma\mathcal{P}_{\text{syst}} = \left(\sum_{s=1}^{S} \left(\sum_{\phi_s = 1}^{\Phi_s} \left(Q_{\text{syst}\,s} - Q_{\text{syst}\,\phi_s}\right)^2 \cdot t_{\phi_s} \right) /8760 \right)^{0.5}$$
$$= \left(\left(\sum_{s=1}^{S} \sigma\mathcal{P}_{\text{syst}\,ms} \right)^2 \cdot T_s/8760 \right)^{0.5};$$

$$\sigma P_{\text{def}\,ms} = \left(\sum_{d=1}^{D_{\text{def}\,md}} \left(P_{\text{def}\,msd} - \text{M}P_{\text{def}\,ms}\right)^2 \cdot q_{\text{def}\,msd} \right)^{0.5};$$

$$\sigma P_{\text{def syst}\,s} = \left(\sum_{d=1}^{D_{\text{def syst}\,s}} \left(P_{\text{def syst}\,sd} - \text{M}P_{\text{def syst}\,s}\right)^2 \cdot q_{\text{def syst}\,sd} \right)^{0.5};$$

$$\sigma P_{\text{def}\,m} = \left(\sum_{d=1}^{D_{\text{def}\,m}} \left(P_{\text{def}\,md} - \text{M}P_{\text{def}\,m}\right)^2 \cdot q_{\text{def}\,md} \right)^{0.5};$$

$$\sigma P_{\text{def syst}} = \left(\sum_{d=1}^{D_{\text{def } m}} \left(P_{\text{def syst } d} - MP_{\text{def syst}} \right)^2 \cdot q_{\text{def syst } d} \right)^{0.5}$$

$$\sigma W_{\text{und } ms} = T_s \cdot \sigma P_{\text{def } ms}; \quad \sigma W_{\text{und syst } s} = T_s \cdot \sigma P_{\text{def syst } s};$$
$$\sigma W_{\text{und } m} = 8760 \cdot \sigma P_{\text{def } m}; \quad \sigma W_{\text{und syst}} = 8760 \cdot \sigma P_{\text{def syst}};$$
$$\sigma \pi_{ms} = \sigma W_{\text{und } ms} / W_{ms}; \quad \sigma \pi_{\text{syst } s} = \sigma W_{\text{und syst } s} / W_{\text{syst } s};$$
$$\sigma \pi_m = \sigma W_{\text{und } m} / W_m; \quad \sigma \pi_{\text{syst}} = \sigma W_{\text{und syst}} / W_{\text{syst}}.$$

In the considered subblock, the summation of the numbers of substantially different exponents (from 10^1 to 10^{-100}) is a serious computational problem. Therefore, it becomes necessary to analyze an exponent of each summand and sum up at the first stage the numbers of close exponents separately and combine the obtained particular sums at the second stage.

The output information includes reliability indices of nodes and the whole system and corresponding root-mean-square deviations, which are determined in the subblock.

Appendix B

B.1 Structure of Methodical Principles for Reliability Accounting in Power Systems Designing

The methodical principles of reliability accounting in power system designing were developed on the basis of the above said in Chap. 3. The principles consist of nine sections and three supplements:

1. Introduction:
   - the purpose of the principles and their sense;
   - the means to support reliability of power supply to consumers;
   - the content of the principles.
2. Application area:
   - for whom (for what organizations) are intended the principles;
   - for the execution of which types of project works, the principles are intended;
   - for the execution of which types of project works, the regulations are intended;
   - which normative and guidance documents should be used along with these regulations.
3. General part—Definitions:
   - short characteristic of the reliability problem as applied to power systems;
   - definition of the main notions used reliability indices.
4. Classification of the design problems solved with reliability accounting:
   - goals and content of the reliability problems, their "place" as applied to design documents;
   - interrelation of individual reliability problems;
   - accepted assumptions and constraints;
   - type of the applied models (optimization or estimation).

G. F. Kovalev and L. M. Lebedeva, *Reliability of Power Systems*,
Power Systems, https://doi.org/10.1007/978-3-030-18736-1

5. Principles of making decisions on development of power systems with accounting of reliability:

   - general scheme of designing;
   - conditions for the development of regulatory recommendations, the use of regulatory requirements for elements, links, and the system as a whole in the development of project solutions;
   - ways for taking into account errors and uncertainty of initial information.

6. Calculated conditions.
   Schemes and calculation states of the system (the modes):

   - the requirements for calculation schemes and for calculation states of systems (the modes) used in solving specific problems;
   - methods and algorithms for the formation and equivalenting of calculation schemes at various levels of the territorial–temporal hierarchy of the control of the development of EPS;
   - choice of the calculated states of the system (modes).

7. Regulatory requirements:

   - "nomenclature" and numerical values of the reliability indices for solving specific problems;
   - regulatory rules and requirements for the structure, parameters, calculating states (the modes) and system elements.

8. Basic principles of solving reliability problems:

   - sequence (stages) and principles of solving the corresponding problems;
   - required initial data;
   - form of representation of the results.

9. Methods for the analysis of the solution results:

   - tasks and principles of the analysis, ways of the analysis;
   - choice of reservation means of different links of system;
   - comprehensive choice of the optimal system development variants.

In addition to the main text, in order to serve as working material for the design, the methodical provisions are supplemented by annexes containing:

- initial data (nomenclature, representation form and if necessary or possible, some numerical values) and representation forms of the results;
- algorithms and programs of methods for solving reliability problems;
- examples of calculations illustrating the use of the regulations (Sects. 5–9).

Note:

1. To Item 6. The methods and algorithms must be presented in the methodical principles, if there are sufficient grounds for recommendations on their application.
2. To Item 9. Principles can be prepared either in a volume covering the solution of only the problem of complex reliability analysis, or in the scope of the problem of system synthesis, taking into account the reliability factor, depending on which of the two forms mentioned in paragraph 3.3, are specific methodical principles.

* * *

In accordance with the suggested structure, "The Recommendations on reliability accounting in power system expansion planning" were issued in 1979 under the aegis of the Scientific Council of the USSR Academy of Sciences for Complex Energy Problems (Moscow, 28 p.).

Appendix C

C.1 Initial Data for Reliability Calculations of the UPS Schemes

This appendix presents the initial data of the eight-unit EEC test circuit. The information is shown in Fig. C.1 and in Tables C.1, C.2, C.3 and C.4. Note that each connecting element consists of one transmission line.

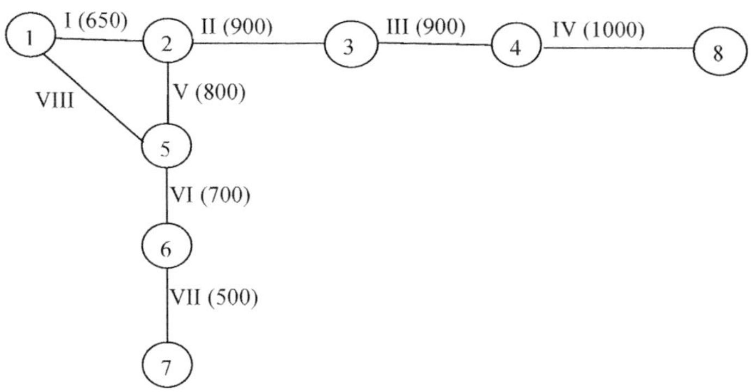

Fig. C.1 Calculation scheme of the test example (the interconnection tie length is given in brackets, km)

G. F. Kovalev and L. M. Lebedeva, *Reliability of Power Systems*, Power Systems, https://doi.org/10.1007/978-3-030-18736-1

Table C.1 Capacity balances by node, MW

Node number (IPS)	Available capacity, P^{g}_{avail}	Repair reserve, P^{g}_{rr}	Working capacity, $P^{g}_{avail} - P^{g}_{rr}$	Peak load, P^{l}_{max}	Emergency reserve, $P^{g}_{emerg\,r}$/%
1	24,750	550	24,200	21,350	2850/13.34
2	39,500	950	38,550	37,750	800/2.12
3	14,950	250	14,700	14,150	550/3.89
4	28,850	650	28,200	27,650	550/1.99
5	45,500	1150	44,350	43,900	450/1.03
6	9550	200	9350	8750	600/6.86
7	10,000	150	9850	7500	2350/31.33
8	9300	200	9100	8100	1000/12.35
UPS	182,400	4100	178,300	169,150	9150/5.4

Table C.2 Composition and parameters of generating units in IPSs and UPS as a whole (working capacities)

Rated capacity of unit, P^g	The failure rate q^g	IPSs																UPS	
		1		2		3		4		5		6		7		8			
		I	$P^g \cdot I$	I	$P^g \cdot I$	I	$P^g \cdot I$	I	$P^g \cdot I$	I	$P^g \cdot I$	I	$P^g \cdot I$	I	$P^g \cdot I$	I	$P^g \cdot I$	I	$P^g \cdot I$
MW	%	pcs.	MW	pcs.	MW	pcs.	MW	pcs.	MW	pcs.	MW	pcs.	MW	pcs.	MW	pcs.	MW	pcs.	MW
50	0.5	62	3100	15	750	21	1050	10	500	51	2550	17	850	53	2650	0	0	229	11,450
50	2.0	100	5000	170	8500	88	4400	170	8500	68	3400	26	1300	33	1650	56	2800	711	35,550
100	0.5	0	0	22	2200	27	2700	11	1100	5	500	0	0	0	0	11	1100	76	7600
100	2.0	18	1800	49	4900	40	4000	31	3100	35	3500	5	500	4	400	16	1600	198	19,800
150	4.5	12	1800	8	1200	1	150	11	1650	6	900	10	1500	17	2550	0	0	65	9750
200	4.5	13	2600	14	2800	12	2400	21	4200	51	10,200	0	0	4	800	0	0	115	23,000
250	0.5	0	0	0	0	0	0	0	0	0	0	4	1000	4	1000	0	0	8	2000
250	4.5	0	0	0	0	0	0	0	0	2	500	0	0	0	0	0	0	2	500
250	5.5	0	0	11	2750	0	0	1	250	2	500	0	0	0	0	0	0	14	3500
300	5.5	15	4500	28	8400	0	0	23	6900	40	12,000	14	4200	0	0	7	2100	127	38,100
350	5.5	0	0	1	350	0	0	0	0	0	0	0	0	0	0	0	0	1	350
400	5.5	0	0	0	0	0	0	0	0	0	0	0	0	2	800	0	0	2	800
450	5.5	2	900	0	0	0	0	0	0	0	0	0	0	0	0	0	0	2	900
450	7.0	0	0	2	900	0	0	0	0	2	900	0	0	0	0	0	0	4	1800
500	4.5	1	500	0	0	0	0	0	0	0	0	0	0	0	0	0	0	1	500
500	7.0	0	0	0	0	0	0	4	2000	0	0	0	0	0	0	3	1500	7	3500
800	7.5	0	0	1	800	0	0	0	0	8	6400	0	0	0	0	0	0	9	7200
1000	7.5	4	4000	3	3000	0	0	0	0	3	3000	0	0	0	0	0	0	10	10,000
1000	9.0	0	0	2	2000	0	0	0	0	0	0	0	0	0	0	0	0	2	2000
Sum		227	24,200	326	38,550	189	14,700	282	28,200	273	44,350	76	9350	117	9850	93	9100	1583	178,300

Note I—amount of units in the group; $P^g \cdot I$—total capacity of units in the group

Table C.3 Conventional load schedules of IPSs (MW/p.u.)

Hour	Numbers of IPSs								UPS
	1	2	3	4	5	6	7	8	
1	21,350	37,750	14,150	27,650	43,900	8750	7500	8100	169,150
	1.0000	1.0000	1.0000	1.0000	1.0000	1.0000	1.0000	1.0000	
2	21,150	37,350	14,100	27,350	43,450	8650	7400	8000	167,350
	0.9906	0.9894	0.9965	0.9892	0.9897	0.9886	0.9867	0.9877	
3	20,950	36,950	13,850	27,050	43,000	8550	7300	7900	165,550
	0.9813	0.9788	0.9788	0.9783	0.9795	0.9771	0.9733	0.9753	
4	20,750	36,550	13,700	26,750	42,550	8450	7200	7800	163,750
	0.9719	0.9682	0.9682	0.9675	0.9692	0.9657	0.9600	0.9630	
5	20,550	36,150	13,550	26,450	42,100	8350	7100	7700	161,950
	0.9625	0.9576	0.9576	0.9566	0.9590	0.9543	0.9467	0.9506	
6	20,350	35,750	13,400	26,150	41,650	8250	7000	7600	160,150
	0.9532	0.9470	0.9470	0.9458	0.9487	0.9429	0.9333	0.9383	
7	20,150	35,350	13,250	25,850	41,200	8150	6900	7500	158,350
	0.9438	0.9364	0.9364	0.9349	0.9385	0.9314	0.9200	0.9259	
8	19,950	34,950	13,100	25,550	40,750	8050	6800	7400	156,550
	0.9344	0.9258	0.9258	0.9241	0.9282	0.9200	0.9067	0.9136	
9	19,750	34,550	12,950	25,250	40,300	7950	6700	7300	154,750
	0.9251	0.9152	0.9152	0.9132	0.9180	0.9086	0.8933	0.9012	
10	19,550	34,150	12,800	24,950	39,850	7850	6600	7200	152,950
	0.9157	0.9046	0.9046	0.9024	0.9077	0.8971	0.8800	0.8889	
11	19,350	33,750	12,650	24,650	39,400	7750	6500	7100	151,150
	0.9063	0.8940	0.8940	0.8915	0.8975	0.8857	0.8667	0.8765	

(continued)

Table C.3 (continued)

Hour	Numbers of IPSs								UPS
	1	2	3	4	5	6	7	8	
12	19,150	33,350	12,500	24,350	38,950	7650	6400	7000	149,350
	0.8970	0.8834	0.8834	0.8807	0.8872	0.8743	0.8533	0.8642	
13	18,950	32,950	12,350	24,050	38,500	7550	6300	6900	147,550
	0.8876	0.8728	0.8728	0.8698	0.8770	0.8629	0.8400	0.8519	
14	18,750	32,550	12,200	23,750	38,050	7450	6200	6800	145,750
	0.8782	0.8623	0.8622	0.8590	0.8667	0.8514	0.8267	0.8395	
15	18,550	32,150	12,050	23,450	37,600	7350	6100	6700	143,950
	0.8689	0.8517	0.8516	0.8481	0.8565	0.8400	0.8133	0.8272	
16	18,350	31,750	11,900	23,150	37,150	7250	6000	6600	142,150
	0.8595	0.8411	0.8410	0.8373	0.8462	0.8286	0.8000	0.8148	
17	18,150	31,350	11,750	22,850	36,700	7150	5900	6500	140,350
	0.8501	0.8305	0.8304	0.8264	0.8360	0.8171	0.7867	0.8025	
18	17,950	30,950	11,600	22,550	36,250	7050	5800	6400	138,550
	0.8407	0.8199	0.8198	0.8156	0.8257	0.8057	0.7733	0.7901	
19	17,750	30,550	11,450	22,250	35,800	6950	5700	6300	136,750
	0.8314	0.8093	0.8092	0.8047	0.8155	0.7943	0.7600	0.7778	
20	17,550	30,150	11,300	21,950	35,350	6850	5600	6200	134,950
	0.8220	0.7987	0.7986	0.7939	0.8052	0.7829	0.7467	0.7654	
21	17,350	29,750	11,150	21,650	34,900	6750	5500	6100	133,150
	0.8126	0.7881	0.7880	0.7830	0.7950	0.7714	0.7333	0.7531	
22	17,150	29,350	11,000	21,350	34,450	6650	5400	6000	131,350
	0.8033	0.7775	0.7774	0.7722	0.7847	0.7600	0.7200	0.7407	

(continued)

Table C.3 (continued)

Hour	Numbers of IPSs								UPS
	1	2	3	4	5	6	7	8	
23	16,950	28,950	10,850	21,050	34,000	6550	5300	5900	129,550
	0.7939	0.7669	0.7668	0.7613	0.7745	0.7486	0.7067	0.7284	
24	16,750	28,550	10,700	20,750	33,550	6450	5200	5800	127,750
	0.7845	0.7563	0.7562	0.7505	0.7642	0.7371	0.6933	0.7160	
Amount	457,200	795,600	298,300	580,800	929,400	182,400	152,400	166,800	–
	21.4145	21.0755	21.0813	21.0054	21.1708	20.8457	20.3200	20.5926	
σ^1, %	3.1	2.3	3.8	2.7	2.2	4.8	5.2	5.0	1.1

Table C.4 Characteristics of ties

Number of tie	Connected nodes	Transfer capabilities of ties						The failure rate of ties	Specific discounted costs, rub./kW
		Variants							
		I	II	III	IV	V	VI		
I	1–2	0	2200	2200	2200	2200	5000	0.0065	6.5
II	2–3	0	3000	3000	3000	3000	5000	0.009	9.0
III	3–4	0	1300	1300	1300	1300	5000	0.009	9.0
IV	4–8	0	400	400	400	1400	5000	0.01	10.0
V	2–5	0	800	1800	1800	1800	5000	0.008	8.0
VI	5–6	0	800	800	1800	1800	5000	0.007	7.0
VII	6–7	0	1000	1000	2000	2000	5000	0.005	10.0
VIII	1–5	0	1000	–	–	–	5000	0.008	8.0

Note

1. The variant with tie 1–5 is calculated additionally only by the program allowing the **ring** ties to be taken into account
2. The variant accounting an emergency rate of ties is calculated additionally only by the programs which take into consideration this factor
3. The specific discounted costs are calculated only in the optimization programs

Appendix D

D.4 Initial Data and Results of Reliability Calculations for Two Variants of the Prospective Structure of the European Part of the Unified Power System (UPS)

See Figs. D.1, D.2, D.3, D.4, D.5, D.6, D.7, D.8, D.9, D.10, D.11, D.12 and D.13; Tables D.1, D.2, D.3, D.4, D.5, D.6, D.7, D.8, D.9, D.10, D.11 and D.12.

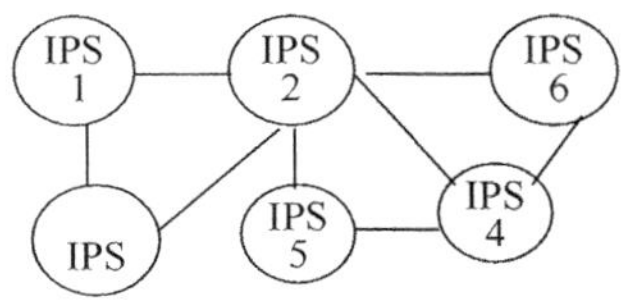

Fig. D.1 Structure of the calculation scheme of a system

G. F. Kovalev and L. M. Lebedeva, *Reliability of Power Systems*, Power Systems, https://doi.org/10.1007/978-3-030-18736-1

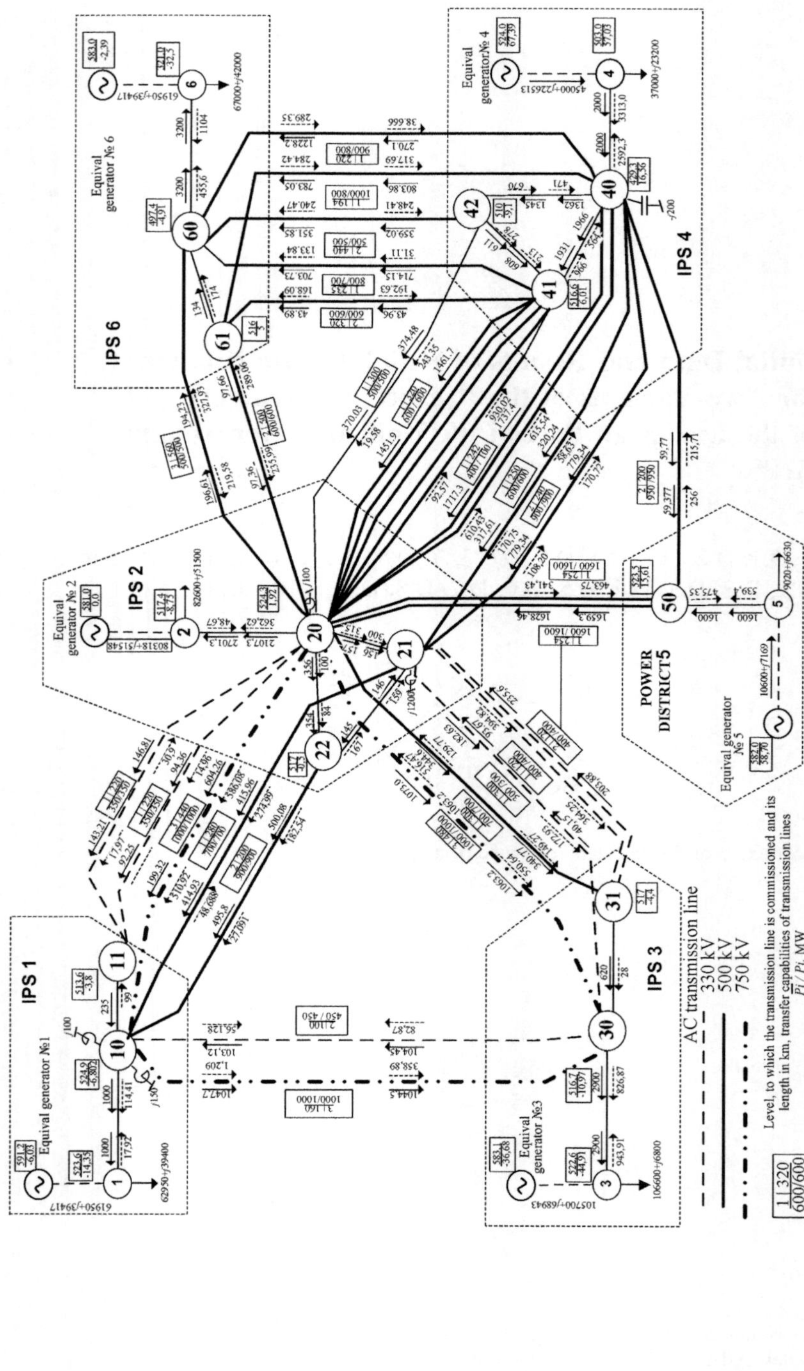

Fig. D.2 Calculation scheme of UPS. Variant 1

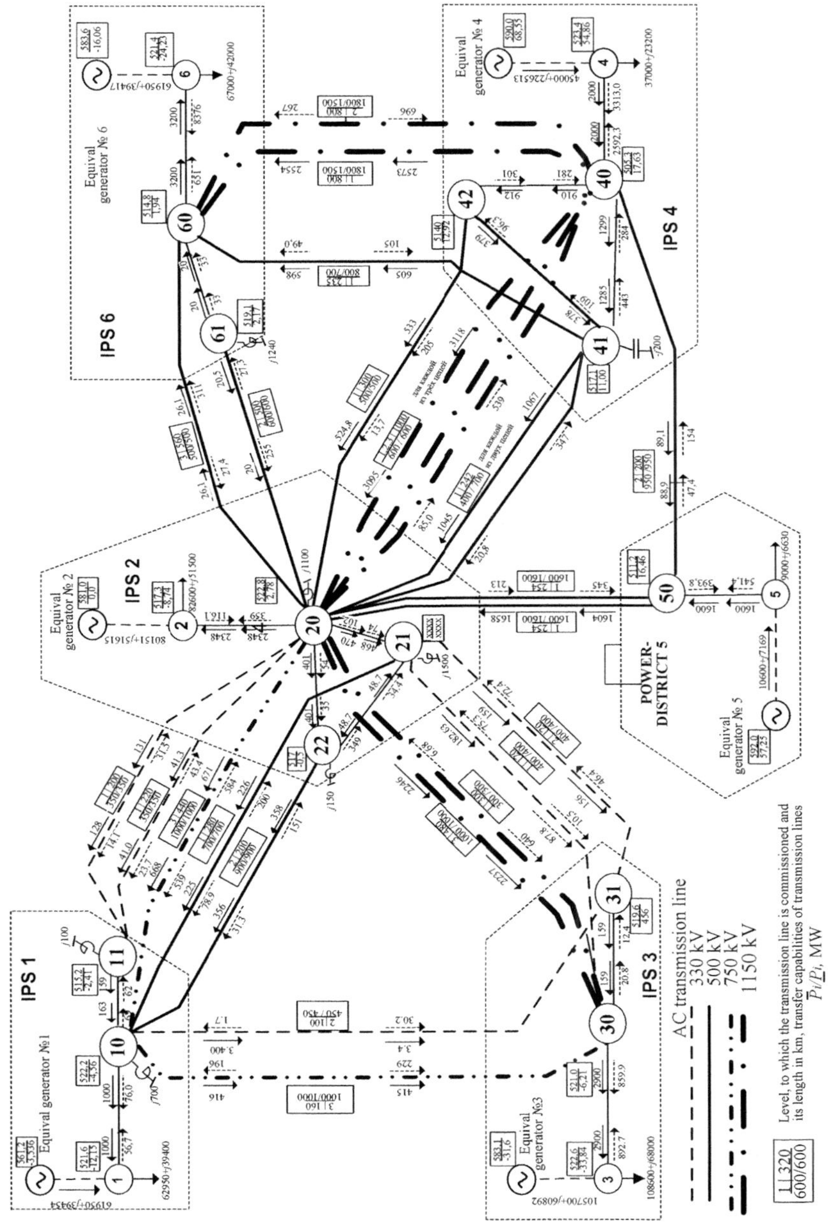

Fig. D.3 Calculation scheme of UPS. Variant 2

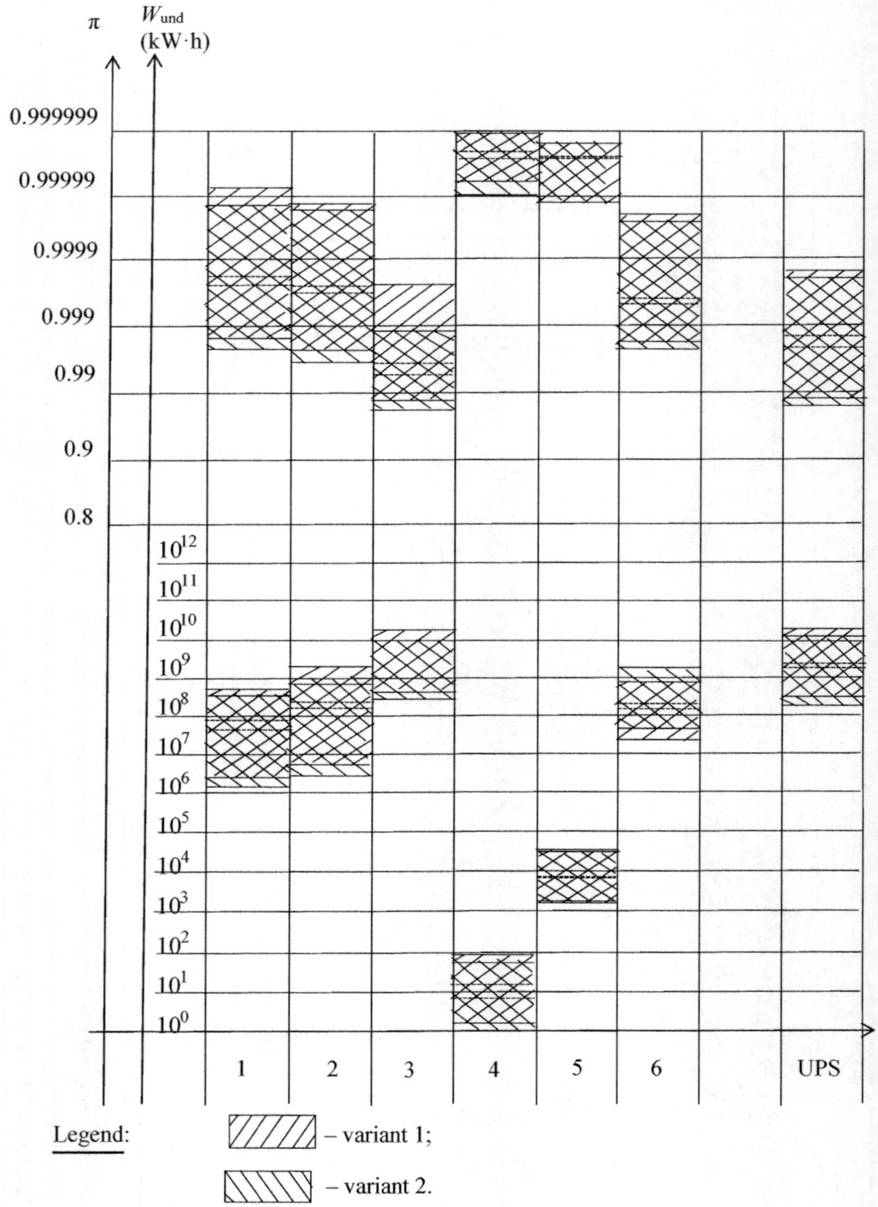

Fig. D.4 Probabilistic ranges of possible values of π and W_{und} for IPSs and UPS, first level

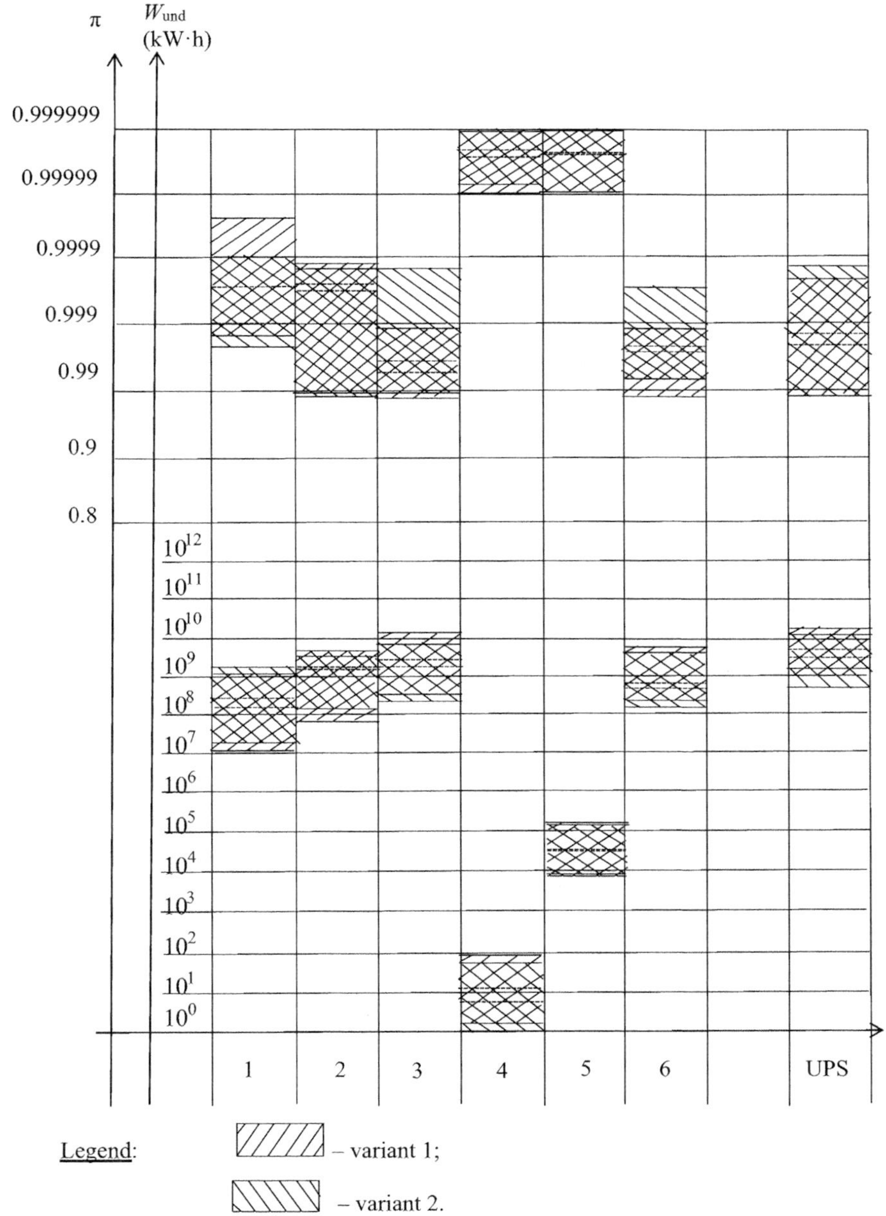

Fig. D.5 Probabilistic values of possible values of π and W_{und} for IPSs and UPS, second level

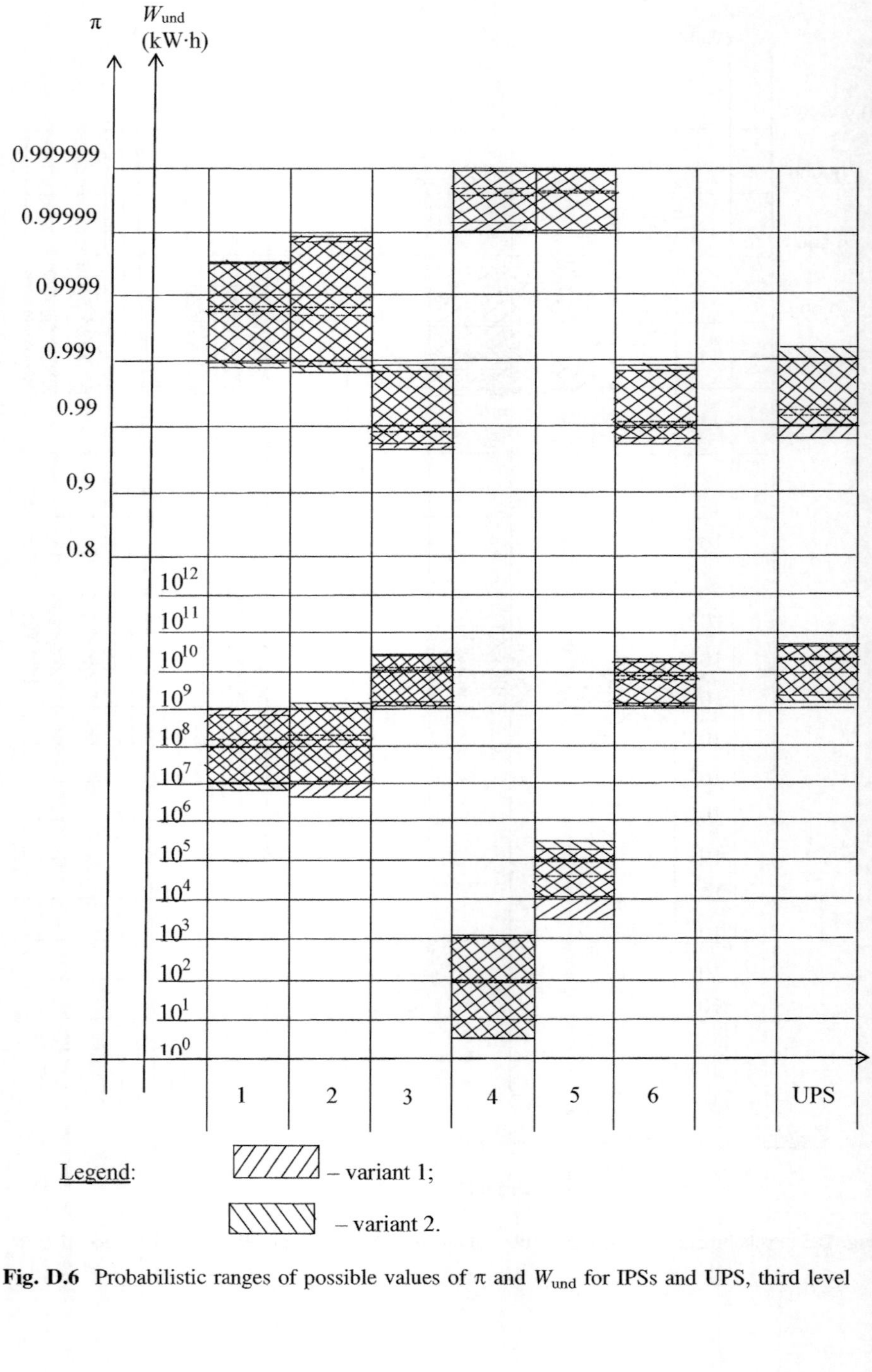

Fig. D.6 Probabilistic ranges of possible values of π and W_{und} for IPSs and UPS, third level

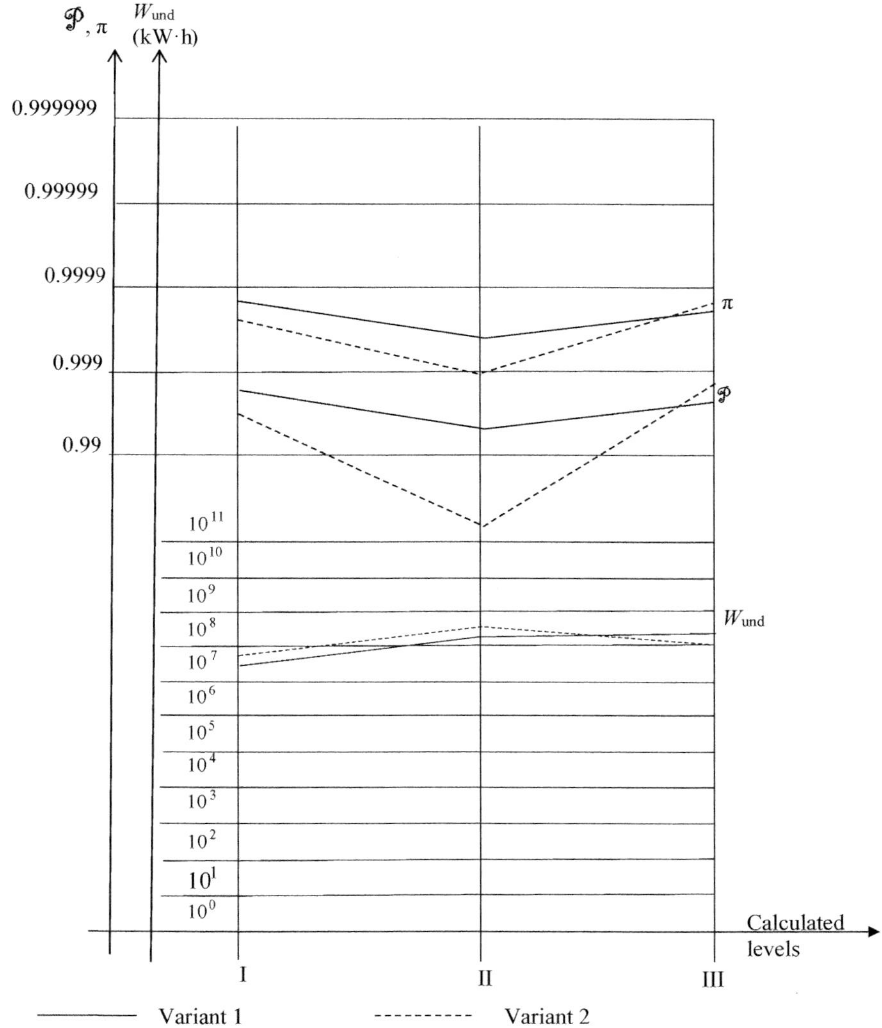

Fig. D.7 Reliability indices of IPS 1 (average values)

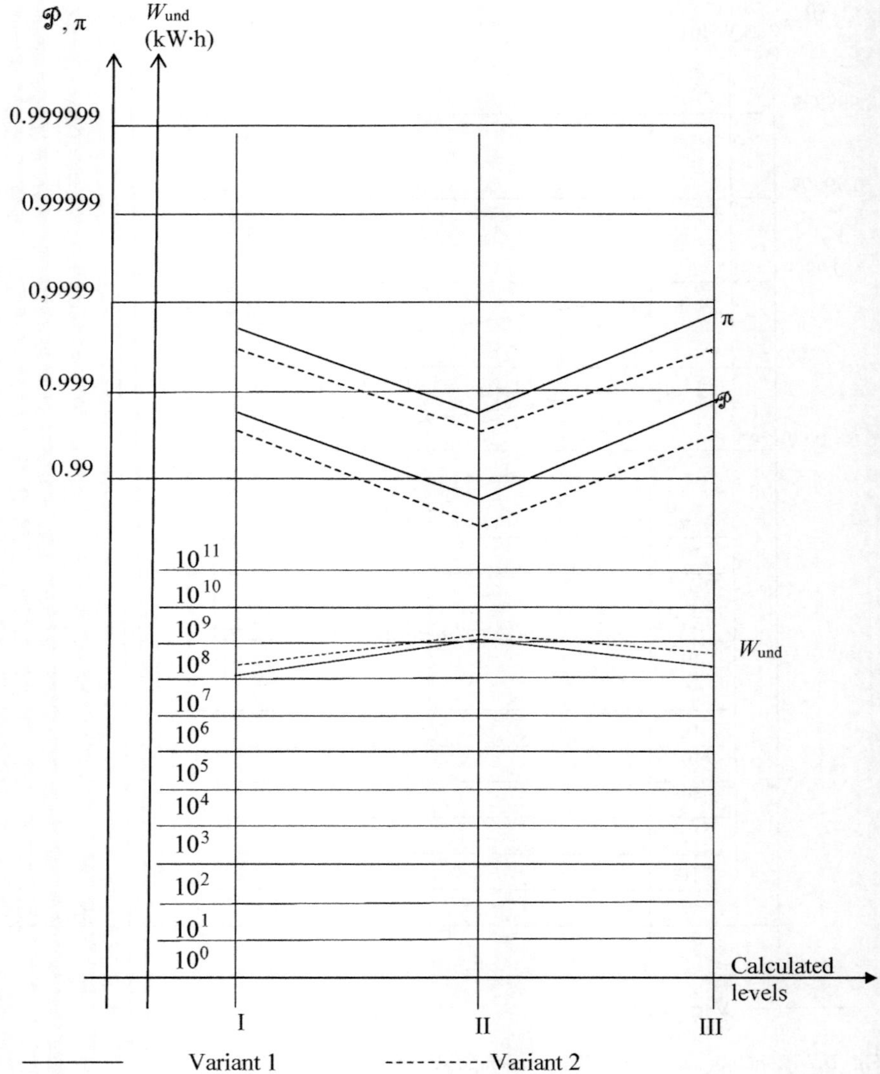

Fig. D.8 Reliability indices of IPS 2

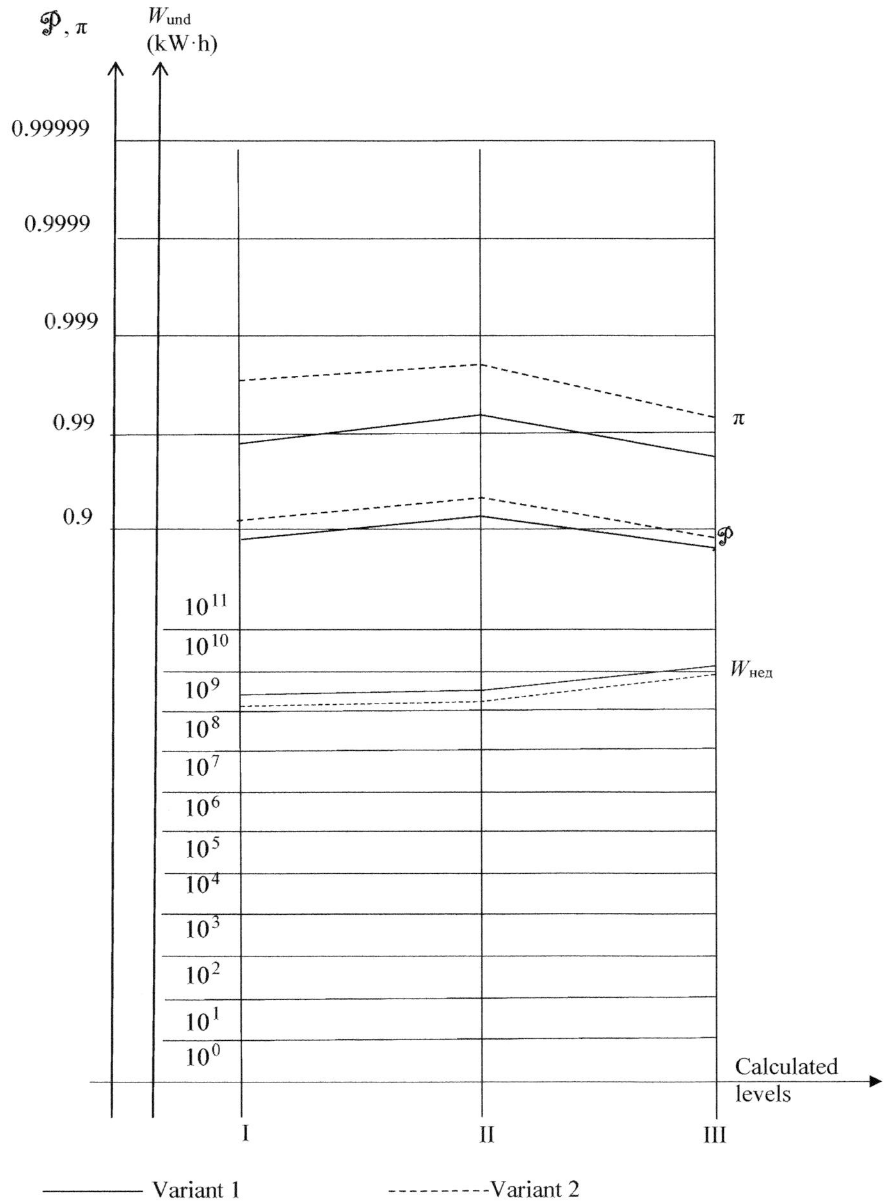

Fig. D.9 Reliability indices of IPS 3

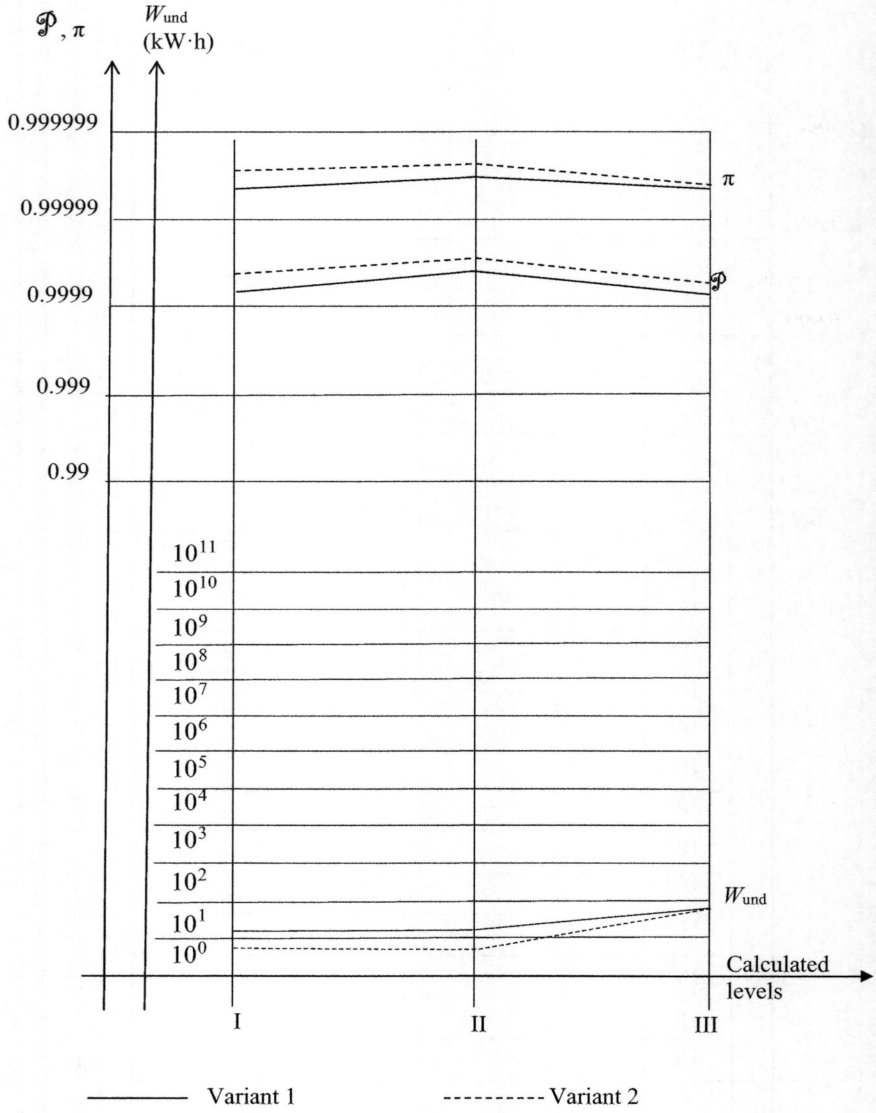

Fig. D.10 Reliability indices of IPS 4

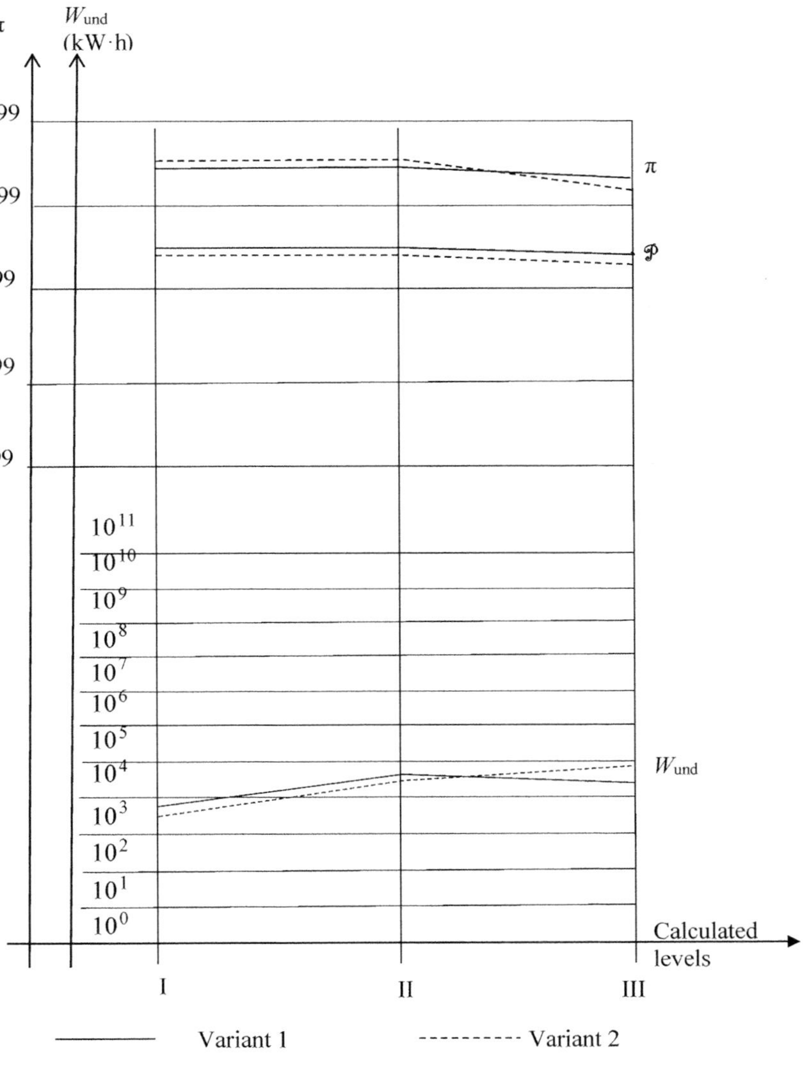

Fig. D.11 Reliability indices of power district 5

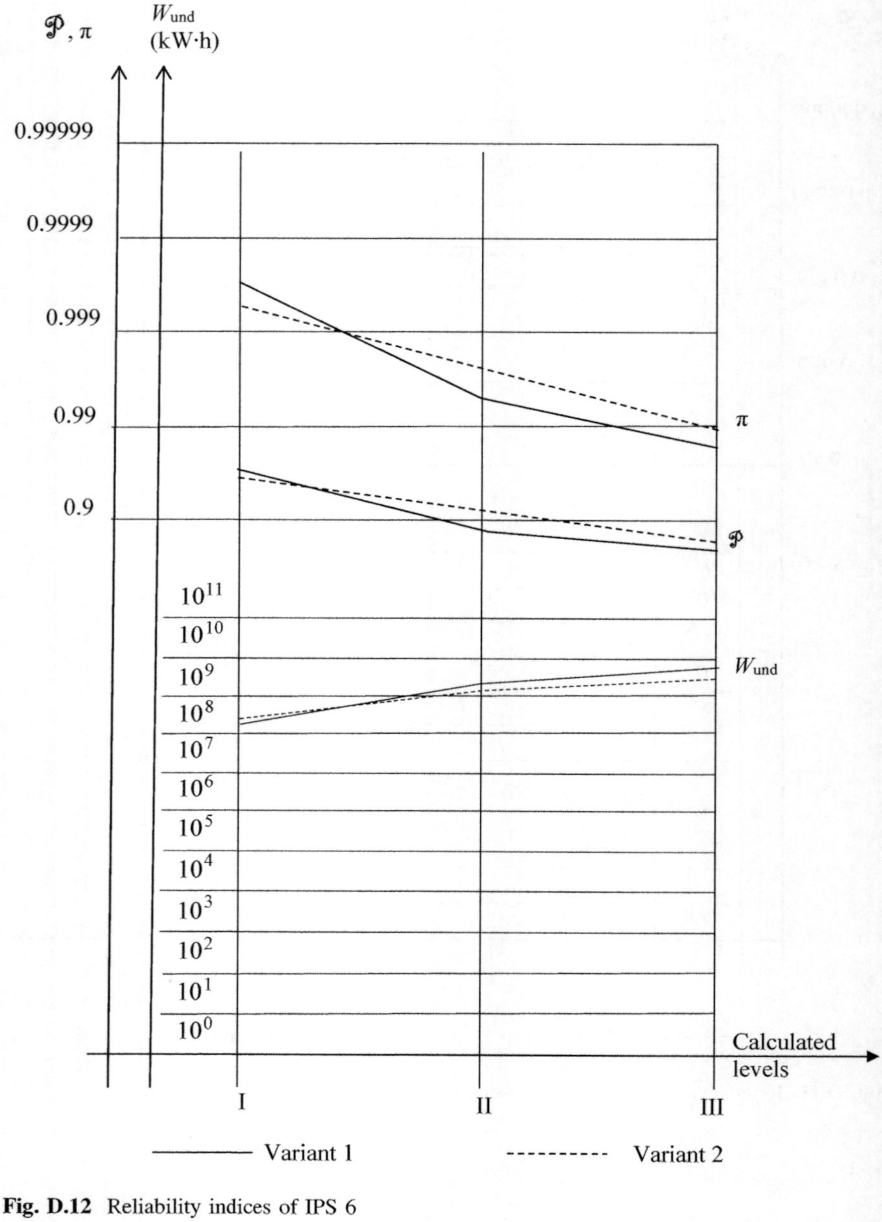

Fig. D.12 Reliability indices of IPS 6

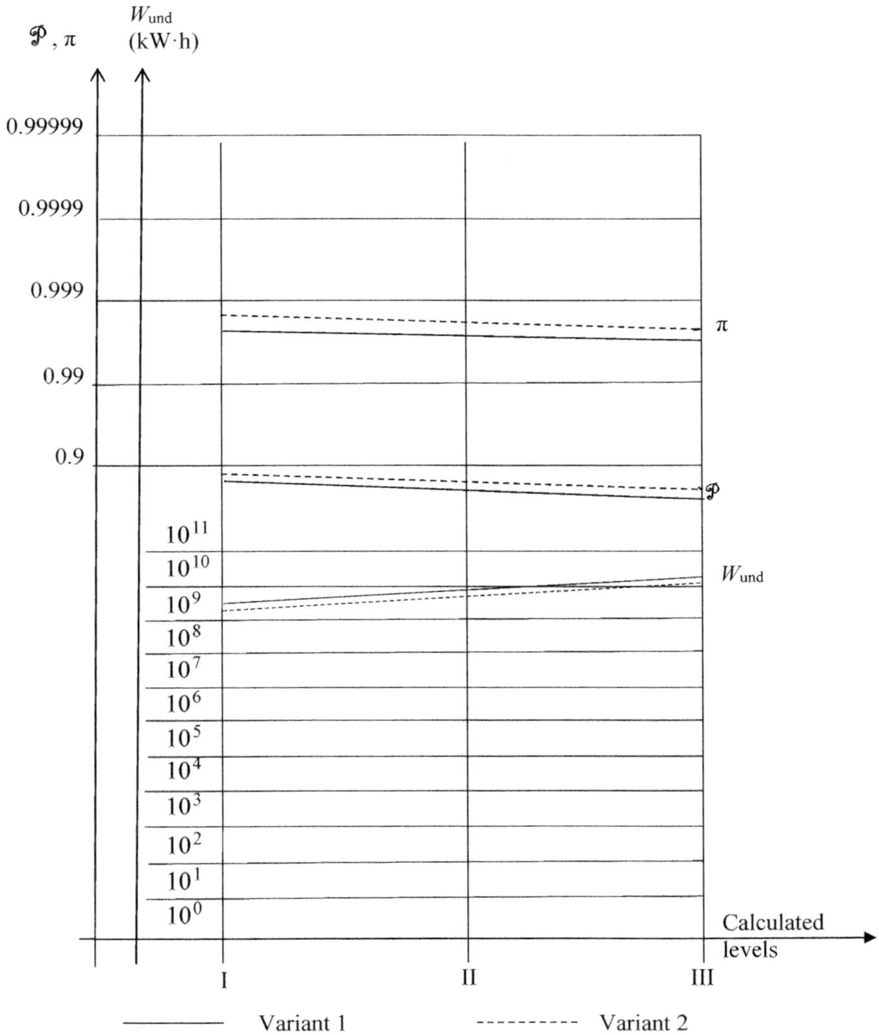

Fig. D.13 Reliability indices of UPS

Table D.1 Absolute peak loads of IPSs, MW

IPSs	Deviation from the Moscow time	Level		
		I	II	III
IPS 1	0	27,300	38,800	54,300
IPS 2	0	38,300	55,600	78,100
IPS 3	0	52,050	74,800	105,400
IPS 4	+1	15,800	23,000	32,600
IPS 5	+1	3950	5350	7300
IPS 6	+2	29,900	42,000	57,100

Table D.2 Annual schedules of monthly peak loads (in % of the absolute peak one)

IPSs	Development level	Months											
		1	2	3	4	5	6	7	8	9	10	11	12
1	I	92.0	89.1	84.4	78.5	72.2	68.2	69.5	72.8	79.2	86.4	93.5	100
	II	93.1	90.1	85.2	79.2	72.8	68.7	69.8	73.0	75.4	86.6	93.6	100
	III	94.3	91.1	86.1	79.9	73.3	69.1	70.1	73.3	79.6	83.7	93.7	100
2	I	93.7	91.6	87.9	83.0	75.3	70.3	71.1	75.5	82.2	88.7	94.7	100
	II	94.5	92.2	88.2	83.0	75.2	70.1	71.0	75.3	81.7	88.5	94.6	100
	III	95.3	92.6	88.7	83.2	75.1	70.0	70.9	75.0	81.3	88.2	94.6	100
3	I	93.5	93.3	90.7	87.1	82.7	79.0	80.1	84.3	88.3	93.4	97.0	100
	II	94.5	93.9	91.1	87.1	82.7	78.8	79.8	8.4	88.5	93.3	97.0	100
	III	95.2	94.6	91.4	87.0	82.8	78.5	79.6	84.5	88.7	94.3	96.9	100
4	I	93.2	91.4	88.4	84.1	77.9	75.5	78.4	81.9	86.7	89.9	95.2	100
	II	94.0	92.0	88.5	83.8	77.1	74.8	77.2	81.2	86.1	89.5	95.0	100
	III	94.3	92.2	88.6	83.6	76.7	74.0	76.9	80.3	85.2	89.1	94.9	100
5	I	92.1	91.9	89.7	86.8	91.5	93.3	95.8	99.2	100	99.0	95.1	97.2
	II	90.2	90.1	87.6	84.7	90.3	91.9	94.5	98.9	100	97.0	93.3	96.1
	III	88.8	88.5	85.8	82.4	89.2	90.9	93.5	97.3	99.4	100	91.7	94.1
6	I	94.8	93.2	90.5	86.5	80.6	77.0	78.0	81.3	86.0	91.0	95.9	100
	II	94.9	93.2	90.3	86.1	80.1	76.6	77.5	80.9	85.6	90.8	95.7	100
	III	95.0	93.2	90.0	85.7	79.6	76.0	77.0	80.3	85.3	90.5	95.5	100

Table D.3 Daily load schedules of IPSs for different levels (I, II, III) in % of $P^{l}_{\max\,\text{daily}}$ (given only for two IPSs—the first and the fourth)

Hours	IPS 1						IPS 4					
	Winter			Summer			Winter			Summer		
	I	II	III	I	II	III	I	II	III	I	II	III
1	57.3	56.6	55.8	61.1	60.5	59.8	63.3	62.3	61.2	69.6	69.1	68.5
2	55.4	55.1	54.9	59.1	58.9	58.8	63.0	62.0	61.0	69.5	68.9	68.4
3	54.9	54.6	54.3	57.8	57.7	57.6	63.0	62.0	61.0	68.9	68.4	67.8
4	54.9	54.6	54.3	57.0	56.9	56.8	65.3	64.2	63.1	69.4	68.8	68.3
5	56.0	55.6	55.2	56.7	56.7	56.6	72.0	70.3	68.6	70.8	70.0	69.1
6	59.4	58.5	57.6	59.2	58.6	58.1	76.8	75.8	74.8	73.5	73.5	73.6
7	67.6	66.8	66.0	65.3	65.0	64.7	91.8	91.1	90.4	90.0	90.4	90.7
8	89.4	88.5	87.5	88.0	87.6	87.1	94.9	94.7	94.6	96.6	97.6	98.6
9	98.2	97.6	97.1	97.4	97.4	97.5	94.6	94.3	94.0	97.8	98.9	100
10	98.5	98.1	97.6	100	100	100	90.5	89.8	89.0	96.5	96.7	97.0
11	94.3	93.2	92.2	98.6	97.6	96.6	86.7	86.5	86.3	92.4	93.2	94.0
12	86.4	86.2	86.0	92.0	92.0	91.9	88.0	87.8	87.7	94.6	95.4	96.2
13	87.9	87.8	87.6	94.1	94.0	94.0	89.0	89.1	89.2	96.5	97.6	98.7
14	90.8	91.0	91.2	96.5	96.7	96.9	88.4	88.1	87.8	94.7	95.5	96.2
15	90.4	90.1	89.8	95.1	94.8	94.5	89.2	88.8	88.4	90.4	91.2	92.1
16	85.3	85.1	85.0	87.0	87.2	87.3	98.0	98.3	98.5	93.3	95.8	98.2
17	92.8	94.3	95.8	89.0	91.5	94.0	100	100	100	94.3	95.8	97.3
18	100	100	100	89.9	90.4	92.9	98.6	98.2	97.8	93.9	94.8	95.8
19	98.2	97.8	97.4	88.4	89.4	90.4	96.2	95.9	95.6	93.6	94.5	95.4
20	95.7	95.4	95.1	87.3	88.2	89.2	94.9	94.5	94.2	97.2	97.7	98.3
21	93.4	93.3	93.1	88.4	89.2	90.0	90.9	89.5	88.1	100	100	98.5
22	88.2	87.1	86.1	90.0	89.2	88.4	80.9	78.6	76.2	91.5	89.6	87.6
23	79.8	77.8	75.8	87.3	85.1	82.8	65.1	63.5	62.0	72.9	71.7	70.6
24	60.6	59.4	58.1	64.6	63.4	62.2	64.0	62.7	61.5	70.9	69.4	69.2

Table D.4 Composition and the failure rate characteristics of generating units (first level) (given for IPSs 1 and 5 only)

No.	Type of power plants	Units capacity, MW	Amount of units	q		
				Optimistic	Pessimistic	Medium
IPS 1						
1.	Block TPPs	1000	2	0.07	0.10	0.085
2.		700	1	0.065	0.095	0.08
3.		500	3	0.06	0.09	0.075
4.		440(R)[a]	2	0.10	0.12	0.11
5.		440	2	0.06	0.09	0.075
6.		300	18	0.05	0.08	0.065
7.		250	5	0.05	0.08	0.065
8.		200	15	0.04	0.08	0.06
9.		160	2	0.04	0.08	0.06
10.		150	7	0.05	0.08	0.06
11.		100	6	0.03	0.05	0.04
12.	Cross-connection TPPs	100	39	0.01	0.03	0.02
13.		50	133	0.01	0.03	0.02
14.	**HPP**	200	1	0.0025	0.0075	0.005
15.		50	84	0.0025	0.0075	0.005
Total		32,530				
etc. for other IPSs						
Power system 5						
1.	Cross-connection TPPs	100	6	0.01	0.03	0.02
2.		50	24	0.01	0.03	0.02
3.	**HPP**	115	22	0.0025	0.0075	0.005
Total		4330				
etc. for other IPSs						

[a]Symbol (R) de. Composition and characteristics of the failure rate of power units (I-st level). Given for nodes 1 and 5 only. Note that this equipment will be in the running-in state at the considered period.

Table D.5 Composition and the failure rate characteristics of generating units (second level) (given for IPSs 1 and 5 only)

No.	Type of power plants	Unit capacity, MW	Amount of units	q		
				Optimistic	Pessimistic	Medium
IPS 1						
1.	Block TPPs	2000(B)	1	0.12	0.14	0.13
2.		1000(B)	4	0.07	0.10	0.085
3.		1000	4	0.07	0.10	0.085
4.		500	9	0.06	0.09	0.075
5.		440	4	0.06	0.09	0.075
6.		300	18	0.05	0.08	0.065
7.		250	12	0.05	0.08	0.065
8.		200	15	0.04	0.08	0.06
9.		160	2	0.04	0.08	0.06
10.		150	6	0.04	0.08	0.06
11.		100	13	0.03	0.05	0.04
12.	Cross-connection TPPs	170	5	0.01	0.03	0.02
13.		100	52	0.01	0.03	0.02
14.		50	143	0.01	0.03	0.02
15.	**HPP**	200	6	0.0025	0.0075	0.005
16.		50	86	0.0025	0.0075	0.005
Total		48,880				
etc. for other IPSs						
Power district 5						
1.	Block TPPs	1000	2	0.11	0.13	0.12
2.	Cross-connection TPPs	100	9	0.01	0.03	0.02
3.		50	22	0.01	0.03	0.02
4.	**HPP**	115	22	0.0025	0.0075	0.005
	Total	6530				
etc. for other IPSs						

Table D.6 Composition and the failure rate characteristics of generating units (third level) (given for IPSs 1 and 5 only)

No.	Type of power plants	Unit capacity, MW	Number of units	q		
				Optimistic	Pessimistic	Medium
IPS 1						
1.	Block TPPs	2000	1	0.07	0.11	0.09
2.		1000	4	0.07	0.10	0.085
3.		1000	10	0.07	0.10	0.085
4.		500	17	0.06	0.09	0.075
5.		440	4	0.06	0.09	0.075
6.		300	18	0.05	0.08	0.065
7.		250	18	0.05	0.08	0.065
8.		200	15	0.04	0.08	0.06
9.		160	2	0.04	0.08	0.06
10.		150	2	0.04	0.08	0.06
11.		100	20	0.03	0.05	0.04
12.	Cross-connection TPPs	170	11	0.01	0.03	0.02
13.		100	64	0.01	0.03	0.02
14.		50	154	0.01	0.03	0.02
15.	**HPP**	200	12	0.0025	0.0075	0.005
16.		50	86	0.0025	0.0075	0.005
Total		65,890				
etc. for other IPSs						
Power district 5						
1.	Block TPPs	1000	1	0.11	0.13	0.12
2.		1000	1	0.07	0.10	0.085
3.	Cross-connection TPPs	100	12	0.01	0.03	0.02
4.		50	22	0.01	0.03	0.02
5.	**HPP**	115	22	0.0025	0.0075	0.005
Total		8830				
etc. for other IPSs						

Table D.7 Standards on current repairs of generating units (in % of the capacity of power generating sets)

Capacity of power generating sets, MW	up to 200	250–300	400–500	600–800	1000 and higher
Reserve value	5.0	5.5	6.0	6.5	7.0

Table D.8 Standards on outages of generating units under major repairs

Capacity of power generating **sets**, MW	up to 100	150–200	250–300	400–500	600–800	1000 and higher
Average duration of major repair (days/year)	15	18	24	28	32	36

Note

1. The table is compiled subject to the major repair works once in 5 years
2. For hydropower units, the major repair works are taken to be once in 6 years with a duration of 15 days

Table D.9 Standards on transmission line outages under scheduled repairs (current and major)

Transmission line voltage, kV	110	230	330	500	750	1150	±400	±750
Average relative duration of scheduled outages	0.0032	0.0047	0.007	0.01	0.012	0.015	0.005	0.007

Table D.10 Probable relative durations of emergency outages of transmission lines (q_{tl})

Transmission line voltage, kV	$q_{tl} = \frac{\tau \sum_{tl}}{T_p}(1/100\,km)$	
	Running-in period (range/average value)	Normal operation (range/average value)
35	0.0015–0.0025/0.002	0.00075–0.00125/0.001
110	0.0015–0.0025/0.002	0.00075–0.00125/0.001
220	0.001875–0.003125/0.0025	0.00075–0.00125/0.001
330	0.001875–0.003125/0.0025	0.00075–0.00125/0.001
500	0.00225–0.00375/0.003	0.00075–0.00125/0.001
750	0.00225–0.00375/0.003	0.00075–0.00125/0.001
1150	0.003–0.005/0.004	0.00075–0.00125/0.001
±400	0.00075–0.00125/0.001	0.000375–0.00625/0.0005
±750	0.00075–0.00125/0.001	0.000375–0.00625/0.0005

Table D.11 Basic parameters of the operating conditions of the IPS, included in the UPS, at various time levels (that were obtained on the basis of Tables D.4, D.5, and D.6)

IPSs	1. Installed capacity of IPSs relative **to** installed capacity of UPS				2. Component of modular capacity relative **to** installed capacity				3. Generating capacity commissioning with respect to P^g installed in IPSs (numerator), modular capacity commissioning with respect to total commissioning (denominator)			4. Ratio between received (plus) or delivered (minus) power of interconnection ties and installed capacity of IPSs		
	Levels				Levels				Levels			Levels		
	Initial	I	II	III	Initial	I	II	III	I	II	III	I	II	III
1	0.15	0.17	0.17	0.17	0.22	0.28	0.40	0.46	$\frac{0.47}{0.36}$	$\frac{0.31}{0.67}$	$\frac{0.30}{0.61}$	+0.055	+0.015	+0.015
2	0.22	0.21	0.22	0.21	0.21	0.30	0.40	0.49	$\frac{0.38}{0.45}$	$\frac{0.32}{0.63}$	$\frac{0.27}{0.71}$	+0.055	+0.050	+0.045
3	0.31	0.32	0.31	0.30	0.44	0.52	0.57	0.67	$\frac{0.42}{0.64}$	$\frac{0.28}{0.68}$	$\frac{0.27}{0.95}$	+0.020	+0.015	+0.010
4	0.09	0.08	0.10	0.11	0	0.07	0.39	0.58	$\frac{0.36}{0.20}$	$\frac{0.46}{0.79}$	$\frac{0.33}{0.96}$	−0.275	−0.240	−0.175
5	0.03	0.04	0.03	0.03	0.41	0.66	0.69	0.78	$\frac{0.42}{1.00}$	$\frac{0.10}{1.00}$	$\frac{0.28}{1.00}$	−0.20	−0.13	−0.14
6	0.20	0.18	0.17	0.18	0.36	0.45	0.57	0.69	$\frac{0.34}{0.61}$	$\frac{0.29}{0.87}$	$\frac{0.30}{0.96}$	+0.05	+0.045	+0.04
UPS	1.0	1.0	1.0	1.0	0.30	0.39	0.49	0.59	$\frac{0.40}{0.52}$	$\frac{0.31}{0.72}$	$\frac{0.29}{0.84}$	–	–	–

Table D.12 Calculation results for IPSs

IPSs	Calculated levels	Total reserve, R	$MP^{g}_{\text{emerg with}}$			$MP^{g}_{\text{emerg without}}$			$\bar{q}_{\text{emerg with}}$			$\bar{q}_{\text{emerg without}}$			$K^{\text{emerg}}_{\text{commis}}$		
			Optimism	Pessimism	Medium	Optimism	Pessimism	Medium	Optimism	Pessimism	Medium	Optimism	Pessimism	Medium	Optimism	Pessimism	Medium
1	I	0.192	935	2433	1685	603	1418	1010	0.027	0.070	0.048	0.017	0.041	0.029	1.55	1.72	1.67
	II	0.260	1672	3521	2600	1045	2330	1688	0.033	0.070	0.052	0.021	0.046	0.034	1.60	1.51	1.54
	III	0.174	2403	4967	3686	1642	3600	2620	0.033	0.069	0.051	0.023	0.050	0.036	1.46	1.38	1.41
2	I	0.122	1416	3471	2445	834	1961	1398	0.032	0.078	0.055	0.019	0.044	0.031	1.70	1.77	1.75
	II	0.049	2331	5368	3850	1418	3303	2310	0.036	0.083	0.059	0.022	0.049	0.036	1.64	1.68	1.67
	III	0.019	3292	7104	5204	2160	4731	3445	0.036	0.079	0.058	0.024	0.053	0.037	1.52	1.50	1.51
3	I	0.191	2762	5774	4271	1649	3562	2606	0.042	0.087	0.064	0.025	0.054	0.039	1.67	1.62	1.64
	II	0.270	3653	7026	5566	2450	5233	3842	0.039	0.076	0.059	0.026	0.056	0.041	1.49	1.34	1.45
	III	0.209	5653	11,296	8473	3754	7854	5805	0.044	0.089	0.067	0.030	0.062	0.046	1.51	1.44	1.46
4	I	0.126	322	981	652	161	436	299	0.019	0.057	0.038	0.009	0.026	0.018	2.00	2.25	2.18
	II	0.162	1228	2548	1891	605	1336	970	0.040	0.083	0.061	0.020	0.043	0.031	2.03	1.91	1.95
	III	0.195	2063	4479	3176	1187	2506	1336	0.045	0.098	0.069	0.026	0.055	0.040	1.74	1.79	1.73
5	I	0.096	389	600	404	202	412	307	0.052	0.080	0.066	0.027	0.055	0.041	1.92	1.46	1.61
	II	0.221	283	540	411	234	475	355	0.034	0.065	0.049	0.028	0.057	0.043	1.20	1.14	1.16
	III	0.210	555	988	771	363	732	547	0.048	0.086	0.067	0.032	0.064	0.048	1.53	1.35	1.41
6	I	0.177	1332	2642	1988	860	1910	1385	0.035	0.070	0.053	0.023	0.051	0.037	1.55	1.38	1.44
	II	0.210	2142	4419	3281	1400	3006	2203	0.041	0.084	0.062	0.027	0.057	0.042	1.53	1.47	1.49
	III	0.190	3600	7425	5513	2282	4773	3528	0.048	0.099	0.073	0.030	0.063	0.047	1.58	1.56	1.57
UPS	I	–	7156	15,901	11,535	4310	9698	7004	0.034	0.076	0.055	0.021	0.047	0.033	1.6	1.64	1.65
	II	–	11,300	23,422	17,600	7153	15,583	11,368	0.038	0.078	0.059	0.024	0.052	0.038	1.58	1.50	1.55
	III	–	17,566	36,259	26,283	11,388	24,195	17,780	0.042	0.086	0.064	0.027	0.057	0.042	1.54	1.50	1.51

Note

1. $MP^{g}_{\text{emerg with}}$—expectation of the capacity under emergency outage with the equipment running-in;
2. $MP^{g}_{\text{emerg without}}$—expectation of the capacity under emergency outage without the equipment running-in;
3. $\bar{q}_{\text{emerg with}}$, $\bar{q}_{\text{emerg without}}$, $K^{\text{emerg}}_{\text{commis}}$ ("commissioning factor")—corresponding values characterizing a relative increase of the capacity under emergency outage caused by the running-in factor of newly commissioned equipment

Appendix E

E.1 Accepted Abbreviations and Notation

The text of this appendix is divided into three parts. The first part expands abbreviations met in the text, the second explains notation which starts with the letters of the Latin alphabet, and the third explains notation which starts with the Greek letters. The sequence of notation corresponds to the sequence of letters in the Latin and Greek alphabets and ending with the Arabian numerals.

I

PSP	Pumped storage plant;
GS	Generator of states;
HPP	Hydropower plant;
UPS	Unified power system;
FFA	Free flow area;
KyrRDE	Kyrgyz Research Department of Energy;
MW	Megawatt;
MW h	Megawatt hour;
p.u.	Per unit
IPS	Interconnected power system;
PC	Personal computer;
RI	Reliability index;
SS	Switching station;
DC	DC link;
CCD	Calculation of capacity deficiency or distribution of capacity deficiency or optimization of capacity deficiency;
DS	Distribution series;
RES	Regional electric power system;
rms deviation	root-mean-square deviation;

(continued)

G. F. Kovalev and L. M. Lebedeva, *Reliability of Power Systems*,
Power Systems, https://doi.org/10.1007/978-3-030-18736-1

(continued)

USA	United States of America;
ⓒ	"Current" (analyzed) calculated system state;
TPP	Thermal power plant;
CP	Co-generation plant
ⓕ	Fixed deficit-free calculated system state;
SRPI	Stock Company G.M. Krzhizhanovsky Power Engineering Institute (State Research Power Institute);
ERCh	Energy reliability characteristic;
EPS	Electric power system.

II

a, A	Current number and amount of the assigned characteristics of daily load curves;
a_{ms}, A_{ms}	Current number and amount of typical daily curves of total load at the m-th node for the s-th interval;
B	Load value measured by the number of units required to serve it;
B_m	Amount of members of the distribution series to simulate irregular load fluctuations;
b_{ims}	Probability of provision with the primary energy resource of the i-th generating element at the m-th node for the s-th interval;
C_I^i	Amount of combinations from I to i;
$C_{I_{ms}}^i$	Amount of combinations from I_{ms} to i;
$C_{I_{ns}}^i$	Amount of combinations from I_{ns} to i;
C_m	Coefficients of the minimized functional of power shortage in the system, which determine the load significance of the m-th node;
$d, D_{\text{def syst}}$	Current number and amount of the distribution series members of power shortages in the system for the whole calculated period T_p;
$d, D_{\text{def syst}\,s}$	Current number and amount of the distribution series members of power shortages in the system for the s-th interval;
$d, D_{\text{def}\,m}$	Current number and amount of the distribution series members of power shortages at the m-th node for the whole calculated period T_p;
$d, D_{\text{def}\,ms}$	Current number and amount of the distribution series members of power shortages at the m-th node for the s-th interval;
$d, D_{ms}(D'_{ms})$	Current number and amount of all possible inconsistent states of generating equipment at the m-th node for the s-th interval;
entire [...]	Integer part of the expression in square brackets;
$F_{\text{valley}\,ms'}$	Valley area in the schedule of monthly peak loads between the peak loads of the s' and $(s'-1)$-th months, which is used for scheduled repairs at the m-th node (MW day);
$F_{\text{valley}\,m(s'-I)}\sum$ $F_{\text{valley}\,ms'}\sum$	Accumulated sums of separate monthly valleys in the m-th node, starting from the month with the lowest monthly peak to $(s'-I)$ and s'-th months (MW day);

(continued)

(continued)

$F_{\text{valley}\,m}$ $F_{\text{valley}\,m\sum}$	Valley area in the schedule of monthly peak loads for the m-th node $F_{\text{valley}\,m\sum} = F_{\text{valley}\,m12\sum}$ (MW day);
$F\left(\bar{P}_{\text{irregular}\,m}\right)$	Distribution function of the irregular load fluctuations of the m-th node, p.u.;
G	Calculated number of the studied system states;
$G_{\text{allowable}}$	Allowable number of calculated system states subject to the limited computing time or volume of RAM
G_s	Calculated number of states at the s-th interval;
h	Value of the last member of the distribution series of generating capacity that remains after its truncation;
H_{in}	length of the i_n-th of transmission line (km);
H_{in}^{equiv}	Length of the i-th transmission line at its equivalent representation;
i, I	Current index and dimension of different sets (system elements, iterations, etc.);
I_j	Amount of generating units in the j-th group;
I_{j_m}	Amount of generating units in the j-th group of the m-th node;
$I_{j_{ms}}$	Amount of generating units in the j-th group of the m-th node for the s-th interval;
i_m, I_m	Current number and amount of units of the m-th node included in the schedule of major repairs;
i_{ms}, I_{ms}	Current number and amount of units (calculated stages of generating capacity) of the m-th node at the s-th interval;
i_n, I_n	Current number and amount of transmission lines in the n-th tie;
i_{ns}, I_{ns}	Current number and amount of transmission lines in the n-th tie at the s-th interval;
I_{limit}	Limiting amount of transmission lines, whose simultaneous outage in one tie is assumed to be quite probable;
j, J_{Gr}	Current number and amount of groups of generators in the system;
j_m, J_m	Current number and amount of groups of units at the m-th node at the beginning of the year;
j_{ms}, J_{ms}	Current number and amount of groups of units at the m-th node at the s-th interval;
k, K	Current index and dimension of the set of different states;
$K_{\text{commiss}}^{\text{accident}}$	Increased emergency rate $\bar{q}_{\text{accident}}$ due to commissioning of new equipment;
k_{filling}^{1}	Load factor of filling schedule;
$K_{\text{screening}}$	Factor of screening of deficit-free states;
K_{valley}	Utilization factor of the valley area in the annual schedule of monthly peak loads for major repairs;
$k_{ds\,s}, K_{ds\,s}$	Current number and amount of members of the ds-th distribution series of system equipment states at the s-th interval;
$k_{ms}\left(k'_{ms}, k''_{ms}\right)$ $K_{ms}(K'_{ms}, K''_{ms})$	Current number and amount of calculated random states of generating capacity at the m-th node of the s-th interval for different equipment sets;

(continued)

(continued)

$k_{ns}\left(k'_{ns}, k''_{ns}\right)$ $K_{ns}\left(K'_{ns}, K''_{ns}\right)$	Current number and amount of calculated random states of transmission lines (calculated values of transfer capabilities) of the n-th tie at the s-th interval for different sets of transmission lines;
K_n^{ERCh}	Amount of members of the energy reliability characteristic of the n-th tie;
k_s, K_s	Current number and amount of calculated random states of the system equipment determined by random values of generating capacities under emergency outage at nodes k_m and random states of transmission lines in ties k_{ns} at the s-th interval;
k_{s0}	Initial value of k_s in the cycle (see the block scheme of the program for calculation of reliability indices of the system);
l, L	Current number and amount of all calculated distribution series of equipment states at the system nodes and in the system ties;
m, M	Current number and amount of nodes (subsystem) in the calculation system scheme;
M^*	Set of system nodes with capacity deficit;
$m', m'', m'''\ldots$	Current numbers of nodes;
max, min	Maximum, minimum;
m_0	Initial value of m in the cycle (see the block scheme of the program for calculation of reliability indices of the system);
$\overline{\text{M}P}{}^{\text{g}}_{\text{emerg}}$	Relative value of mathematical expectation of the capacity being in the emergency state;
$\text{M}P^{\text{g}}_{\text{emerg without}}$	Expected value of the capacity being in the outage state without equipment running-in;
$\text{M}P^{\text{g}}_{\text{emerg with}}$	Mathematical expectation of the capacity being in the emergency outage with equipment running-in;
$\text{M}P_{ms}$, $\text{M}P'_{ms}$, $\text{M}P''_{ms}$	Mathematical expectation of the generating capacity being in the fault-free state at the m-th node at the s-th interval for different equipment compositions, respectively;
$\text{M}P_{\text{def syst}}$	Mathematical expectation of capacity deficit in the system during the year;
$\text{M}P_{\text{def}\,m}$	Mathematical expectation of capacity deficit at the m-th node during the year;
$\text{M}P_{\text{def syst}\,s}$	Mathematical expectation of capacity deficit in the system at the s-th interval;
$\text{M}P_{\text{def}\,ms}$	Mathematical expectation of capacity deficit at the m-th node at the s-th interval;
$\text{M}W_{\text{und}}$, $\text{M}W'_{\text{und}}$	Mathematical expectation of electricity undersupply to consumers;
$\text{M}W_{\text{und syst}}$	Mathematical expectation of electricity undersupply to consumers in the system during the period T_{p};
$\text{M}W_{\text{und}\,m}$	Mathematical expectation of electricity undersupply to consumers at the m-th node during the year;
$\text{M}W_{\text{und syst}\,s}$	Mathematical expectation of electricity undersupply to consumers in the system at the s-th interval;

(continued)

(continued)

$MW_{\text{und}\,ms}$	Mathematical expectation of electricity undersupply to consumers at the m-th node at the s-th interval;
n, N	Current number and amount of ties in the calculation system scheme;
n_m, N_m	Current number and amount of ties adjacent to the m-th node;
N^*	Amount of ties, for which ERCh is determined;
$n0$	Initial value of n in the cycle (see the block scheme of the "YANTAR" program for the calculation of system reliability indices);
n_{equi}	Amount of equivalent generators;
$P^{\text{g}}_{\text{emerg r}}$	Emergency reserve of generating capacity;
$P^{\text{g}}_{\text{outage}\,k_{ms}}$	Calculated k_{ms}-th value of the total generating capacity of the m-th node at the s-th interval, which is in emergency outage;
$P^{\text{g}}_{\text{unit}}$	Generating unit capacity;
$P_{\text{def syst}\,d}$	Capacity deficit in the d-th system state;
$P_{\text{def syst}\,sd}$	Capacity deficit in the d-th system state at the s-th interval;
$P_{\text{def syst}\,\varphi_s\beta k_s}$	Capacity deficit in the $\varphi_s\beta k_s$-th system state;
$P_{\text{def}\,md}$	Capacity deficit at the m-th node for the d-th state;
$P_{\text{def}\,msd}$	Capacity deficit value at the m-th node at the s-th interval for the d-th state;
$P_{\text{def}\,m\,\varphi_s\beta k_s}$	Capacity deficit at the m-th node in the $\varphi_s\beta k_s$-th system state;
$P_{\text{def step syst}}$	Assigned capacity step for distribution series of capacity deficit in the system;
$P_{\text{def step}\,m}$	Assigned capacity step for distribution series of capacity deficit at the m-th node;
$P^{\text{g}}_{\text{major r}m}$	Calculated value of the required capacity reserve for major repairs at the m-th node;
$P^{\text{g}}_{\text{major r}\,ms(\chi)}$	Calculated value of the capacity for major repairs of generating equipment at the m-th node at the $s(\chi)$-th interval;
$P^{\text{g}}_{\text{major r}\,m\varphi_s}$	Calculated value of the capacity for major repairs of generating equipment at the m-th node for the φ_s-th calculated subinterval;
$P^{\text{l}}_{\text{monthly max}\,ms}$	Monthly peak loads at the m-th node at the s-th, s'-th, $(s'+1)$-th, and χ-th intervals, respectively;
$P^{\text{l}}_{\text{irreg}\,m\beta}$, $P^{\text{l}}_{\text{month max}\,ms'}$, $P^{\text{l}}_{\text{month max}\,m(s'+1)}$, $P^{\text{l}}_{\text{month max}\,m\chi}$	Calculated (β-th) values of the irregular load component at the m-th node;
$\bar{P}_{\text{irreg}\,\beta_m}$	β_m-th step of the irregular load fluctuation of the m-th node, p.u.;
P_{opr}	Calculated (for the peak load period) value of the operating reserve;
$P^{\text{g}}_{\text{planed r}\,ms}$	Total capacity of units of the m-th node, which are in scheduled repairs at s-th an interval;
$P^{\text{g}}_{\text{work}}$	Operating generating capacity;
P^{g}_{rr}	Capacity of repairs' reserve;
$P^{\text{g}}_{\text{avail}\,m}$	Available generating capacity of the m-th node;
$P^{\text{g}}_{\text{avail}\,ms}$	Available generating capacity of the m-th node for the s-th interval;

(continued)

(continued)

$P^{\mathrm{l}}_{\mathrm{comb\,max}}$	Combined absolute peak of load of system;
$P^{\mathrm{l}}_{\mathrm{comb\,conv\,max}}$	Combined absolute peak of conventional load of the system;
$P^{\mathrm{l}}_{\mathrm{comb\,conv\,min}}$	Combined minimum of conventional load of the system;
$P^{\mathrm{l}}_{\mathrm{average}}$	Average load value;
$P^{\mathrm{l}}_{\mathrm{average}\,m}$	Average load value of the m-th node;
$P^{\mathrm{l}}_{\mathrm{day\,max}\,ms\alpha}$	Peak of the α-th daily load schedule of the m-th node at the s-th interval;
$P^{\mathrm{l}}_{\mathrm{requi}\,m}$	Required consumed capacity at the m-th node;
$P^{\mathrm{l}}_{\mathrm{cur\,r}\,ms}$	Calculated value of the capacity for current repairs at the m-th node at the s-th interval;
$P^{\mathrm{g}}_{\mathrm{instal}}$	Installed capacity;
$P^{\mathrm{g}}_{\mathrm{instal}\,m}$	Installed generating capacity at the m-th node;
$P^{\mathrm{g}}_{\mathrm{instal}\,ms}$	Installed generating capacity at the m-th node for the s-th interval;
$P_{\mathrm{step}\,m}, P'_{\mathrm{step}\,m}, P''_{\mathrm{step}\,m}$	Calculated steps of capacity change or distribution series of states of the generating equipment of the m-th node;
$P_{\mathrm{step}\,n}$	Calculated steps of capacity change for distribution series of states of the n-th tie and for its transmission lines;
P_{equi}	Equivalent generator capacity;
P^{g}_{d}	Generating capacity in the d-th state of the node or system;
P^{l}_{d}	Load in the d-th state of the node or system;
P^{g}_{i}	Capacity of the i-th generating aggregate;
$P^{\mathrm{g}}_{i_m}$	Rated (or available) capacity of the i-th unit at the m-th node;
$P^{\mathrm{g}}_{i_{ms}}$	Rated (or available) capacity of the i-th unit at the m-th node at the s-th interval;
P^{g}_{j}	Generator capacity of the j-th group;
$\underline{P}_{i_{nt}}, \bar{P}_{i_{nt}}$	Transfer capability limits of the n-th transmission line in the t-th tie in the directions taken as negative and positive (inverse and direct), respectively;
p_{i_n}	Probability of failure-free operation of the n-th transmission line in the t-th tie;
$\underline{P}_{i_{ns}}, \bar{P}_{i_{ns}}$	Transfer capability limits of the n-th transmission line in the t-th tie at the s-th interval in the directions taken as negative and positive (inverse and direct), respectively;
$p_{i_{ns}}$	Probability of failure-free operation of the i-th transmission line in the n-th tie at the s-th interval;
$P^{\mathrm{g}}_{j_m}$	Available capacity of the unit in the j-th group at the m-th node;
$P^{\mathrm{g}}_{j_{ms}}$	Available capacity of the unit in the j-th group at the m-th node at the s-th interval;
$p_{j_{ms}}$	Probability of failure-free operation of the unit in the j-th group at the m-th node at the s-th interval;
$P^{\mathrm{tl}}_{k}, P^{\mathrm{tl}}_{k'}, P^{\mathrm{tl}}_{k''}$	k, k' and k''-th states of transmission lines in the ties;
$P^{\mathrm{g}}_{k_{ms}}, P^{\mathrm{g}}_{k'_{ms}}, P^{\mathrm{g}}_{k''_{ms}}$	Values of generating capacities in the k, k' and k''-th states for different equipment sets at the m-th node at the s-th interval;

(continued)

(continued)

$\underline{P}_{k_{ns}}, \bar{P}_{k_{ns}}$, $\underline{P'}_{k_{ns}}, \bar{P'}_{k_{ns}}$, $\underline{P''}_{k_{ns}}, \bar{P''}_{k_{ns}}$	Transfer capability limits of the n-th tie in its k, k' and k''-th states at the s-th interval in the inverse and direct directions for different composition of transmission lines;
$P^{\text{tl}}_{k_{ns}}, P^{\text{tl}}_{k'_{ns}}, P^{\text{tl}}_{k''_{ns}}$	k, k' and k''-th states of different sets of transmission lines in the n-th tie at the s-th interval;
P^{g}_{m}	Calculated value of the consumed generating capacity at the m-th node;
P^{l}_{m}	The calculated value of the capacity consumption consumed in the m-th node;
$P^{\text{l}}_{\max}$	Peak load capacity;
$P^{\text{l}}_{\max m}$	Absolute peak load of the m-th node;
$P^{\text{lconv}}_{\max m}$	Absolute peak of the conventional load of the m-th node;
$P^{\text{g}}_{\min i_m}$	Least capacity of the i-th unit at the m-th node;
$P^{\text{equiv}}_{m'm''}, P^{\text{equiv}}_{m'm'''}$, $P^{\text{equiv}}_{m''m'''}$,	Equivalent transfer capabilities of transmission lines between the corresponding nodes $m', m'', m''', \ldots$;
P^{l}_{msat}	Load at the m-th node in the t-th hour of the a-th daily schedule at the s-th interval;
$P^{\text{l conv}}_{msat}, P^{\text{l conv}}_{msat^*_s}$	Conventional (calculated) load at the m-th node of the a-th daily schedule of the s-th interval for the t-th and t^*_s-th hours;
$P^{\text{l}}_{m\varphi_s}$	Regular (average) load of the m-th node at the φ_s-th subinterval of the s-th interval (in terms of auxiliary power consumption and losses in distribution networks);
$P^{\text{l conv}}_{m\varphi_s}$	Regular (average) conventional load of the m-th node at the φ_s-th subinterval of the s-th interval (in terms of auxiliary power consumption and losses in distribution networks);
$P^{\text{l}}_{m\varphi_s\beta}$	The required load in the m-th node in the φ_s-th state (mode) taking into account the irregular fluctuation β;
$P^{\text{l conv}}_{m\varphi_s\beta}$	The required conventional load of the m-th node in the φ_s-th state (mode) taking into account irregular fluctuations β;
$P^{\text{g}}_{m\varphi_s\beta k_s}$	The generating capacity used at the m-th node in the $\varphi_s\beta k_s$-th state (mode);
$P^{\text{l}}_{m\varphi_s\beta k_s}$	The served load at the m-th node in the $\varphi_s\beta k_s$-th state (mode);
$P^{\text{l conv}}_{m\varphi_s\beta k_s}$	The served conventional load at the m-th node in the $\varphi_s\beta k_s$-th state (mode);
$\underline{P}_n, \bar{P}_n$	Maximum transfer capabilities of the n-tie in the inverse and direct directions;
P^{eq}_{n}	Transfer capability of the equivalent n-th tie;
$P_{n\varphi_s\beta k_s}$	Capacity flow in the n-th tie in the $\varphi_s\beta k_s$-th state (capacity flow from this node to neighboring ones is taken here for positive direction, from the adjacent nodes to this one—for negative direction);
$P^{\text{lconv}}_{\sum sat}, P^{\text{lconv}}_{\sum sat^*_s}$	Total conventional load of the t-th and t^*_s-th hours of the a-th daily schedule in the system at the s-th interval;
$P^{\text{lconv}}_{\sum \varphi_s}$	Same for the φ-th subinterval of the s-th interval;

(continued)

(continued)

$\mathscr{P}$, $\mathscr{P}_{g}$	(a) probability of failure-free power supply; (b) availability;
$\mathscr{P}_{tu}$	Steady-state availability factor;
$\mathscr{P}_{syst}$	Availability factor of the system during the period T_p;
$\mathscr{P}_{syst\,s}$	Availability factor of the system for the s-th interval;
$\mathscr{P}_{m}$	Availability factor of the system for the m-th node during the period T_p;
$\mathscr{P}_{ms}$	Availability factor of the system for the m-th node at the s-th interval;
q	Probability (probable relative duration) of outage of the system element;
$\bar{q}_{emerg}$	Averaged probability of emergency outage of available capacity;
$\bar{q}_{emerg\,no-run}$	Averaged probability of emergency outage without equipment running-in;
$\bar{q}_{emerg\,run}$	Averaged probability of emergency outage of available capacity with equipment running-in;
$q_{defsyst\,d}$	Probability of the d-th capacity deficit state (mode) of the system;
$q_{def\,syst\,sd}$	Probability of the d-th capacity deficit state (mode) of the system at the s-th interval;
$q_{def\,syst\,\varphi_s\beta k_s}$	Probability of capacity deficit state (mode) $P_{def\,syst\,\varphi_s\beta k_s}$ in the system;
$q_{def\,md}$	Probability of the d-th capacity deficit state of the m-th node during the period T_p;
$q_{def\,msd}$	Probability of the d-th capacity deficit state (mode) of the m-th node at the s-th interval;
$q_{def\,m\varphi_s\beta k_s}$	Probability of capacity deficit $P_{def\,m\varphi_s\beta k_s}$ at the m-th node at the s-th interval in the $\varphi_s\beta k_s$-th state (mode);
q_{tl}	Probability of transmission line outage;
$q_{tl\,f}$	Factual probability of transmission line outage;
q_{opt}	Optimistic estimation of the probability of equipment failure (outage);
q_{ever}	Average probability value of equipment failure (outage);
q_{pessim}	Pessimistic estimation of the probability of equipment failure (outage);
$q_{norm\,i}$	Probability of emergency outage of the i-th unit during its normal operation;
$q_{run-in\,i}$	Probability of emergency outage of the i-th unit during the running-in period;
Q	Probability of failure state (mode);
Q_{syst}	Probability of deficient state (mode) of the system during the period during the period T_p;
$Q_{syst\,s}$	Probability of deficient state (mode) of the system at the s-th interval;
$Q_{syst\,\varphi_s}$	Probability of deficient states (modes) of the system at the φ_s-th calculated subinterval of the s-th interval;
Q_m	Probability of deficient states (modes) of the system because of deficient state of the m-th node during the year;
Q_{ms}	Probability of deficient states (modes) of the m-th node at the s-th interval;
$Q_{m\varphi_s}$	Probability of deficient states (modes) of the m-th node at the φ_s-th calculated subinterval of the s-th interval;
$q_d\left[P_d^g\right]$	Probability of the d-th state (mode) of generating capacity P_d^g;

(continued)

(continued)

q_i^{g}	Probability of emergency outage of the i-th unit;
$q_{i_{ms}}$	Probability of emergency outage of the i-th unit at the m-th node at the s-th interval;
q_{i_n}	Probability of outage of the i-th transmission line in the n-th tie;
$q_{i_{ns}}$	Probability of outage of the i-th transmission line in the n-th tie at the s-th interval;
q_j^{g}	Probability of emergency outage of units in the j-th group;
$q_{j_{ms}}^{g}$	Probability of emergency outage of units in the j-th group of the m-th node at the s-th interval;
$q_k\left[P_k^{g}\right]$, $q_{k+1}\left[P_{(k+1)}^{g}\right]$	Probability of the k-th and $(k + 1)$-th states (modes) of generating capacities P_k^{g} and $P_{(k+1)}^{g}$;
$q_{k_{ds}}$	Probability of the k-th member of the ds-th distribution series of system equipment states (modes) at the s-th interval;
$q_{k_{ms}}\left[P_{k_{ms}}^{g}\right]$, $q_{k'_{ms}}\left[P_{k'_{ms}}^{g}\right]$, $q_{k''_{ms}}\left[P_{k''_{ms}}^{g}\right]$	Probability of the k-th, k'-th, and k''-th calculated states of generating capacities $P_{k_{ms}}^{g}, P_{k'_{ms}}^{g}, P_{k''_{ms}}^{g}$ at the m-th node at the s-th interval;
$q_{k_{ns}}$	Probability of the k-th calculated state of the transmission line in the n-th tie at the s-th interval;
$q_{k_{ns}}\left[P_{k_{ns}}^{tl}\right]$, $\left(q_d\left[P_d^{tl}\right]\right)$ $q_{k'_{ns}}\left[P_{k'_{ns}}^{tl}\right]$, $q_{k''_{ns}}\left[P_{k''_{ns}}^{tl}\right]$	Probability of the k-th (d-th), k'-th, and k''-th states of transmission lines in the n-th tie, for which their total transfer capabilities are equal to $P_{k_{ns}}^{tl}\left(P_d^{g}\right), P_{k'_{ns}}^{tl}, P_{k''_{ns}}^{tl}$, respectively;
q_{k_s}	Probability of the k-th system equipment state at the s-th interval;
q_m^{g}	Averaged probability of emergency outage of generating equipment at the m-th node during the calculated period T_p;
q_{ms}^{g}	Averaged probability of emergency outage of generating equipment at the m-th node at the s-th interval;
q_{syst}^{g}	Averaged probability of outage of generating equipment in the system during the calculated period T_p;
q_{equiv}	Emergency rate of equivalent generator;
q_n^{equiv}	Probability of outage of the equivalent n-th tie;
$q_{n\,permiss}^{ERCh}$	Probability of the n-th tie operation with any possible admissible capacity flow;
$q_{n\,permiss'}^{ERCh}$	Probability of the n-th tie operation with one of the calculated capacity flows;
q_{β_m}	Probability of the β-th calculated random deviation of load of the m-th node;
$q_{\varphi_{s\beta}}$	Probability of the m-th random irregular deviation of system load at φ_s-th subinterval;

(continued)

(continued)

q_{0i_n}	Specific probability (per 100 km of length) of outages of the *i*-th transmission line in the *n*-th tie in emergency and scheduled repairs;
$q_{0i_n}^{\text{equiv}}$	Specific probability (per 100 km of length) of outages of the *i*-th equivalent transmission line in the *n*-th tie in emergency and scheduled repairs;
$q_{0i_{ns}}$	Specific probability (per 100 km of length) of outage of the *i*-th transmission line in the *n*-th tie at the *s*-th interval in emergency and scheduled repairs;
$q_{0i_{ns}}^{\text{equiv}}$	Same for equivalent transmission line;
$q_{0i_{ns}}^{\text{emerg}}$	Specific probability (per 100 km of length) of outage of the *i*-th transmission line in the *n*-th tie in emergency repairs at the *s*-th interval;
$q_{0i_{ns}}^{\text{schedule}}$	Specific probability (per 100 km of length) of outage of the *i*-th transmission line in the *n*-th tie in scheduled repairs at the *s*-th interval;
r_{op}	Relative value of operating reserve;
$r_{\text{reserv}\sum}$	Relative value of total calculated reserve;
R	Amount of units comprising operating reserve;
$RS1$	Logical variable that determines availability of fixed deficit-free state (mode) of the system;
$RS2$	Logical variable that determines possibility of brief analysis of the current system state (mode) only on the basis of the analysis of its equipment state;
$RS3$	Logical variable that determines possibility of brief analysis of the current system state (mode) only on the basis of irregular load deviations at the nodes;
s, S	Current number and amount of the calculated intervals in the calculated period T_{p};
s'	Current number of the ranked sequence of months in the year;
$s0$	Initial value of *s* in the cycle (see the block scheme of the program "YANTAR" for calculation of the system reliability indices);
t	Current number of time intervals; number of hour of the daily load schedule;
$\bar{t}_{\text{major r}}$	Standard of scheduled major and medium repairs of system equipment;
$\bar{t}_{\text{major r}\,i_m}$	Standard of scheduled major and medium repairs of the *i*-th unit at the *m*-th node (day/year);
$\bar{t}_{\text{major r}\,j_m}$	Standard of scheduled major and medium repairs for units of the *j*-th group at the *m*-th node;
T_{p}	Calculated period taken equal to one year (8760 h);
T_{calc}	Computing time of reliability indices of the researched system by the computer program;
t_{calc}	Computing time of one system state by the computer program;
T_j	Interval of full operability of the system;
$T_{\sum}$	Total duration of failure-free operation for the period T_{p};
$T_{\max\,\text{syst}}^{\text{l}}$	Amount of utilization hours peak load of the system;
$T_{\max\,m}^{\text{l}}$	Amount of utilization hours peak load of the *m*-th node;

(continued)

(continued)

T_s	Duration of the s-th interval;
t_s^*	Set of hours t, whose conventional (calculated) load at each of M nodes of the s-th interval is fixed;
t_{φ_s}	Duration of the φ-th subinterval at the s-th interval;
$V_{\text{major r}\,m}$	Required of major repairs of generating equipment at the m-th node (MW day);
W'_m	Right-hand sides of linear algebraic equations $m = \overline{1, M}$;
W	Required power generation;
W_{fact}	Factual power generation;
W_{syst}	Required power generation in the system for the period T_{p};
$W_{\text{syst}}^{\text{conv}}$	Volume of power corresponding to the conventional load schedule of the system for the period T_{p};
$W_{\text{syst}\,s}$	Power generation in the system at the s-th interval;
W_m	Required power generation at the m-th node for the period T_{p};
W_m^{conv}	Annual volume of power corresponding to the conventional load schedule of the m-th node;
W_{ms}	Power generation at the m-th node for the s-th interval;
W_{und}	Average value of power undersupply for a certain period;

III

$\bar{\alpha}_{\text{curr}\,i_{ms}}$	Standard (relative total duration) of scheduled current repairs of the i_m-th unit at the m-th node at the s-th interval;
$\bar{\alpha}_{\text{curr}\,j_{ms}}$	Standard (relative total duration) of scheduled current repairs for the unit in the j-th group at the m-th node at the s-th interval;
β, B	Current number and amount of calculated random values of irregular load components in the system;
β_m, B_m	Current number and amount of calculated random values of irregular load components at m-th node;
β_0	Initial value of β in the cycles of the program "YANTAR" for calculation of reliability indices;
γ, Γ	Current number and amount of all system cutsets;
$\delta_{i_{ns}k_{ns}}$	Variable taking the value of 1 or 0 depending on the operability of the i-th transmission line in the k-th state of the n-th tie at the s-th interval;
δ_{1n}	Functional variable $\Psi, n = \overline{1, N}$;
δ_{2m}	Functional variable $\Psi, m = \overline{1, M}$;
$\Delta_{k_{ns}}^{\text{tl}}$	Set **of** $\delta_{i_{ns}k_{ns}}$ for all transmission lines in the k-th state of the n-th tie at the s-th interval;
ΔP_j	Failure level, MW;
$\Delta P^{\text{l}}_{\text{syst}\,\varphi_s\beta k_s}$	Minimized value of the not covered load in the system as a whole in the $\varphi_s\beta k_s$-th state (mode);
$\Delta P^{\text{opt}}_{\text{syst}\,\varphi_s\beta k_s}$	Optimal (minimized) value of the non-served load $\Delta P^{\text{l}}_{\text{syst}\,\varphi_s\beta k_s}$ in the system in the $\varphi_s\beta k_s$-th state (mode)of the s-th interval;
$\Delta P^{\text{g}}_{m\varphi_s\beta k_s}$	Excess of generating capacity at the m-th node in the $\varphi_s\beta k_s$-th state (mode);

(continued)

(continued)

$\Delta P^{\mathrm{l}}_{m\varphi_s\beta k_s}$	Minimized value of the non-served not covered load at the *m*-th node in the $\varphi_s\beta k_s$-th state (mode);
$\Delta P^{\mathrm{opt}}_{m\varphi_s\beta k_s}$	Optimal (minimized) value of the the non-served load $\Delta P^{\mathrm{l}}_{m\varphi_s\beta k_s}$ at the *m*-th node in the $\varphi_s\beta k_s$-th state (mode);
$\Delta\pi$	Calculated error in determination of the index π;
ε_q	Calculation accuracy of distribution series of generating capacity;
ε_r	Calculation accuracy of balance equations;
ε_Ψ	Calculation accuracy of the auxiliary minimized functional Ψ;
$\eta_{i_{ns}k_{ns}}$	Variable taking the value of 1 or $p_{i_{ns}}/q_{i_{ns}}$ depending on the value of $\delta_{i_{ns}k_{ns}}$;
$v_{a_{ms}}$	Number of days for the *m*-th node at the *s*-th interval based on the a_{ms}-th daily schedule of load;
v_{st}	Number of *t*-th operating hours of the system with the same loads at nodes in the s-th interval;
π	Energy provision factor;
π_{syst}	Same for the system;
$\pi_{\mathrm{syst}\,s}$	Same for the *s*-th interval;
π_m	Same for the *m*-th node during the period T_{p};
π_{ms}	Same for the *s*-th interval;
$\bar{\rho}$	Error of peak load forecast;
ρ_{ms}	The value of the peak load $P^{\mathrm{l}}_{\mathrm{month\,max}\,ms}$ *m*-th node in the *s*-th interval in fractions of $P^{\mathrm{l}}_{\mathrm{max}\;ms}$, taken as 1.0;
ρ_{msa}	Value of the daily peak load $P^{\mathrm{l}}_{\mathrm{dayly\,max}\,msa}$ of the *a*-th schedule of the *m*-th node at the *s*-th interval in fractions of $P^{\mathrm{l}}_{\mathrm{month\,max}\,ms}$ taken equal to 1.0;
ρ_{msat}	Value of the peak load P^{l}_{msat} of the *t*-th hour of the *a*-th daily schedule of the *m*-th node at the *s*-th interval in fractions of $P^{\mathrm{l}}_{\mathrm{dayly\,max}\,msa}$ taken equal to 1.0;
σ^{l}	rms deviation of irregular load fluctuations;
σ^{l}_m	rms deviation of irregular load fluctuations of the *m*-th node;
$\sigma P'_{ms}, \sigma P''_{ms}, \sigma P_{ms}$	rms deviation of the failure-free state of the generating capacity of different composition of units of the *m*-th node at the *s*-th interval;
$\sigma \mathrm{P}_{\mathrm{syst}}$	rms deviation of the availability factor of the system for the calculated period T_{p};
$\sigma \mathrm{P}_{\mathrm{syst}\,s}$	The same for the *s*-th interval;
$\sigma \mathrm{P}_m$	The same for the *m*-th node;
$\sigma \mathrm{P}_{ms}$	The same for the *s*-th interval;
$\sigma P_{\mathrm{def\,syst}}$	rms deviation of deficit capacity in the system for the calculated period T_{p};
$\sigma P_{\mathrm{def\,syst}\,s}$	The same for the *s*-th interval;
$\sigma P_{\mathrm{def}\,m}$	rms deviation of deficit capacity at the *m*-th node for the calculated period T_{p};
$\sigma P_{\mathrm{def}\,ms}$	The same for the *s*-th interval;
$\sigma W_{\mathrm{und\,syst}}$	rms deviation of power undersupply in the system during the calculated period T_{p};
$\sigma W_{\mathrm{und\,syst}\,s}$	The same for the *s*-th interval;

(continued)

(continued)

$\sigma W_{\text{und}\,m}$	rms deviation of power undersupply at the m-th node during the calculated period T_{p};
$\sigma W_{\text{und}\,ms}$	The same for the s-th interval;
$\sigma \pi_{\text{syst}}$	rms deviation of the coefficient of provision by energy of consumers in the system for the calculated period T_{p};
$\sigma \pi_{\text{syst}\,s}$	The same for the s-th interval;
$\sigma \pi_m$	rms deviation of the coefficient of provision by energy of consumers of the m-th node for the period T_{p};
$\sigma \pi_{ms}$	The same for the s-th interval;
$\tau_{i_{ms}\sum}$	Mathematical expectation of the total duration of emergency outages of the i_m-th unit during the period T_s;
τ_j	Interval of the j-th system failure state;
φ, Φ	Current number and amount of calculated subintervals during the period T_{p};
φ_s, Φ_s	Current number amount of calculated subintervals at the s-th interval that are determined by the fixed values of the regular load $P^{\text{l}}_{m\varphi_s}$ at all nodes;
φ_{s0}	Initial value of φ_s in the cycles (see the block scheme of the program "YANTAR" for calculation of system reliability indices);
χ	Current number of the calculated intervals (see s) or the month (see s')

References

1. Recommendations on accounting for reliability in the design of the development of electric power systems (1973) Academy of Sciences of the USSR, Scientific Council on Complex Energy Problems, Moscow, 28 p (in Russia)
2. Lyalik GN, Urvantsev NP (1972) Determination of the emergency power reserve based on universal characteristics of the specific reserve, № 1, pp 26–30 (in Russia)
3. Volkenau IM, Dubrovskaya RA, Kovalev GF, Rudenko YN (1973) On the account of reliability in the design of electric power systems. In: Sat. methods for calculating the emergency reserve in complex power systems. Proceedings of the Kirghiz Research Department of Energy, issue 4. Publishing House “Kirghizstan”, Frunze, pp 1–7 (in Russia)
4. Markovich IM (1969) Modes of energy systems. Energia, Moscow, 352 p (in Russia)

G. F. Kovalev and L. M. Lebedeva, *Reliability of Power Systems*,
Power Systems, https://doi.org/10.1007/978-3-030-18736-1

Printed by Printforce, the Netherlands